MADE AT HOME
PRESERVES

Thank you to everyone who has taken the time to teach us so that we can now pass it on to Indy and anyone else who will listen

Preserves
by Dick & James Strawbridge

First published in Great Britain in 2012
by Mitchell Beazley, an imprint of
Octopus Publishing Group Limited,
Endeavour House, 189 Shaftesbury Avenue,
London, WC2H 8JY
www.octopusbooks.co.uk

An Hachette UK Company
www.hachette.co.uk

ISBN: 978-1-84533-658-5

A CIP catalogue record for this book is available from the British Library

Printed and bound in China

Neither the authors nor the publishers take any responsibility for any injury or damage resulting from the use of techniques shown or described in this book. Advice on natural remedies in this book is not intended as a substitute for medical advice and use of them is entirely at the reader's own risk.

Both metric and imperial measurements are given for the recipes. Use one set of measures only, not a mixture of both.

Standard level spoon measurements are used in all recipes
1 tablespoon = 15ml
1 teaspoon = 5ml

Ovens should be preheated to the specified temperature. If using a fan-assisted oven, follow the manufacturer's instructions for adjusting the time and temperature. Grills should also be preheated.

This book includes dishes made with nuts and nut derivatives. It is advisable for those with known allergic reactions to nuts and nut derivatives and those who may be potentially vulnerable to these allergies, such as pregnant and nursing mothers, invalids, the elderly, babies and children, to avoid dishes made with nuts and nut oils.

It is also prudent to check the labels of preprepared ingredients for the possible inclusion of nut derivatives.

The Department of Health advises that eggs should not be consumed raw. This book contains some dishes made with raw or lightly cooked eggs. It is prudent for more vulnerable people such as pregnant and nursing mothers, invalids, the elderly, babies and young children to avoid uncooked or lightly cooked dishes made with eggs.

MADE AT HOME

DICK & JAMES STRAWBRIDGE

PRESERVES

MITCHELL BEAZLEY

CONTENTS

INTRODUCTION

There is nothing quite like a shelf or a cupboard that is brimming with home-made produce. Not all of us can grow or rear our own, but it is possible to make full use of seasonal food to store for the rest of the year. It wasn't so long ago that every household had the skills to preserve and store its food – indeed it was a matter of life or death to ensure you had enough for your family to survive. We are fortunate enough to live in a time where our lives no longer revolve around providing food for the table, so we have passed responsibility to others for growing, rearing, finding, storing and even cooking our food. This convenience should not be undervalued as it allows us time for leisure – and the 'good old days' were undoubtedly hard.

For many people, it is becoming increasingly important to know what they are eating and drinking: additives are treated with suspicion, so it's logical to want to make your own produce. That being said, let's not forget the enjoyment that comes from putting delicious, homemade food on the table for friends and family. We are all used to mass-produced preserves that by their very nature have a consistent flavour, texture and appearance. Commercial producers have to ensure that their customers get what they expect every time, so all individuality is removed. With homemade preserves it is all but impossible to achieve exactly the same result every time. This variation is to be celebrated, and you may well find that the last jar of a favourite chutney made several years ago will be closely guarded and used in a positively miserly manner by your family. But don't worry, when it does finally disappear, another, probably very different, chutney will take over the mantle of being number one.

When you have a small selection of your own preserves at home you will find it very simple to transform even the simplest meal or snack. A homemade cake filled with your own jam and a little cream, or simple lunch of bread, cheese or pâté and your own pickle or chutney will be savoured because of all the effort that has been invested in the meal, albeit months in advance. If you're new to preserving, the tricky part is choosing where to begin and which methods to learn, and of course that will all depend on what your favourite preserves are. For us it's quite simple – we have very eclectic tastes, so will try anything and everything.

SOURCING YOUR FOOD

It goes without saying that the better your fresh ingredients are, the better your preserves will be. The good news is that this doesn't mean you've got to track down special sources of exotic foods, or spend a fortune buying them. While it's true that supermarkets bring us every possible type of produce all year round, you will find that local seasonal food is undoubtedly superior. When you decide to preserve, it makes no sense to put all that effort into anything but the best quality you can lay your hands on, as it's very simply a matter of the final product reflecting what went in. True, flavours are enhanced and modified by the addition of sugar, salt, vinegars and spices, or are concentrated by drying. However, once you've opened that jar, it's no good wishing you'd done something different six months earlier.

There is also the matter of price: produce is cheapest when it is at its most abundant. It's a win/win situation – you get the best choice and you pay less, so it would seem sensible to make the most of this great opportunity. If you grow your own – and of course we recommend that you do – you will have the happy problem of having to deal with gluts. There is always some point when the abundance of tomatoes, courgettes, apples, plums, gooseberries, strawberries or raspberries will mean you're not sure if you want to eat another, even though they are truly fantastic. These are the best moments to invest a little time in preserving the harvest for use at any time of year.

SKILLS AND FLAVOURS

It is very important to remember that there are lots of possible variations when it comes to the methods and recipes in this book. Once you understand the principles, you can vary the flavours. You will have the joy of taking full credit for the magnificent preserves you produce, so feel free to fine-tune the recipes and make them your own.

In the following pages we show you how to make the most of local, seasonal produce when it is at its very best and most affordable. In return for a few hours' work, your homemade preserves will brighten up countless meals throughout the year, as well as making wonderful gifts. All you have to do is decide where you would like to begin, and you will soon find your store cupboard filling up with jarred, dried and bottled treats.

Dick & James

PREPARING TO

PRESERVE

Wondering where to begin? The great thing about preserves is the variety of products you can make, so the decision is all about what you really enjoy. Without a doubt you should start by making something you love the taste of. If you have a sweet tooth, you could go for the old faithful of making strawberry jam, but you may do well to head a little out of your comfort zone and try a fruit cheese. They may not be as popular as jams, and the texture is slightly different, but a quince or damson cheese is a delight and the intensity of the fruit flavour is a wonderful thing. If you love your savouries, a chutney or a pickle is a great route to achieving satisfaction – however, you'll need to wait a while for the flavours to meld together and for your endeavours to reach their optimum.

STORAGE

Once you start making preserves, you will have the task of storing your creations – and it doesn't take long for you to amass shelves of jars. Space can be at a premium, and pantries tend to be only just big enough for storing the weekly shopping and the storeroom staples we take for granted. If you don't have much space in your kitchen, clean out an old cupboard or find a shelf in your shed. Anywhere that's clean, dry and sheltered will do. Organisation is important too, especially if you've stored your preserves elsewhere. We label our jars carefully, date all of our preserves and arrange our shelves by produce. You might believe you will remember exactly what you have made and where you've put it, but after six or nine months or even more, it is not surprising to forget. Keep a set of labels in one of your empty jars (you will find you'll begin to collect a supply) so they are always to hand when needed.

FRESH PRODUCE

It may seem unusual to have a section on storing fresh produce in a book on preserving. However, it is important to store food in a way that preserves the flavours and textures for as long as possible. Some vegetables store extremely well and it is not necessary to process them prior to keeping them – it is more a matter of careful selection, a slight clean and then putting them in the right environment to stop them degrading.

It is all well and good telling you that most produce is best stored in a dark, dry and cool location. In many homes, every effort is made to keep all areas light and warm, so you might need to find somewhere outside, unless you are fortunate enough to have an unheated cellar. The challenge of storing underground is to keep the area dry, but the upside is that the temperature stays pretty constant. Any place that is protected from the sun's heat and from frost in the winter – and, of course, draughts at any time of year – will make a great fresh produce store.

Outbuildings (or in our case an old outside privy) that have a good roof and are well insulated, with windows that can be covered, make great stores. Sadly, this rules out most garden sheds, which tend to heat up with a little sunshine.

DRYING

Removing the moisture from produce is a preserving technique that has been used for millennia. The flavours are simply concentrated, and this is drying's strength. Anyone can preserve this way, and hanging herbs in your kitchen is a simple starting point. You can dry fruit, vegetables and spices outside in the fresh air, in a homemade solar dryer (see pages 38–41) or in an oven (fan-assisted ovens are particularly good), or you can use a food dehydrator, a small electrical appliance with a heating element and fan.

JARS

Most of us have cupboards and fridges full of jars. If you're going to make your own jarred goodies, you will have to buy – or better yet, collect – a supply of jars to fill. There is a plethora of containers that can be bought specifically for preserving, lots of reusable Kilner jars, both clip-top and screw-lid, as well as artisan jam jars, and no shortage of bargain jars, be they round, square or even hexagonal. Jars from our everyday shopping can be cleaned and reused. It's a little annoying if the lids are branded, but as any jar sizes are standardised, you will probably be able to find a plain lid that fits. Of course, storing jars in anticipation of filling them with preserves requires space. We have tried to keep ours in the garage or a shed but somehow they always get dusty and lids migrate away from the jars, so we now keep our supply in a cupboard in the kitchen or utility room.

As far as essential kit goes, any large pan can be used for preserving, though there are special preserving pans available for those who intend to do a lot. Steer well clear of aluminium and invest in stainless steel – we picked up ours second hand. A preserving funnel is extremely useful as it has a wide neck to allow thick chutneys or jams to flow through and it stops a lot of mess and wastage. Buy one or make your own from the top of a plastic bottle.

BOTTLES

Learning to bottle your own syrups, sauces and alcoholic drinks is a natural

progression from making your own jam or chutney. It's the cordials, syrups and ketchups that are particularly satisfying to produce: you will have a collection of very handsome bottles that can be used at will, and they also make great presents. Yet again, it is extremely worthwhile to collect bottles for future use – especially elegant ones and those with ornamental stoppers.

The flavours of your home-made bottled treats will be unique, and of course not exactly like those of commercially available products. This is particularly true when it comes to making your own cider, beer and wine, either from grapes or fruit, flowers or herbs, but we consider the taste of our home-made alcoholic drinks to be well worth acquiring, and get through gallons of them.

Brewing and fermentation equipment is essential, so start collecting well in advance of when you intend to start. Masticators and presses can be shared or borrowed, so ask around and you could save yourself the cost. Stock up on bottle brushes and 'magic balls' – little ball bearings that you swish around in the bottom of a glass container to scour away any dirt.

FREEZING

Not everything can be frozen, but freezing is a simple, effective and popular method of preserving produce. Despite this, few people are actually good at it, as retaining the integrity and texture of meat, fruit or vegetables requires using the correct techniques. Most households have a domestic freezer or at least a freezer compartment in their fridge, so you will require very little else to get started. The key is organization: if you don't use what you have put in the freezer while it is still at its best, you might as well not have bothered to freeze it. Label your freezer containers or bags effectively using a decent marker pen – write down the date as well as the contents – and make sure to visit your freezer regularly to use the goodies you have squirrelled away. Any other items you need, such as pans for blanching or trays on which to freeze produce, can be found in the average kitchen.

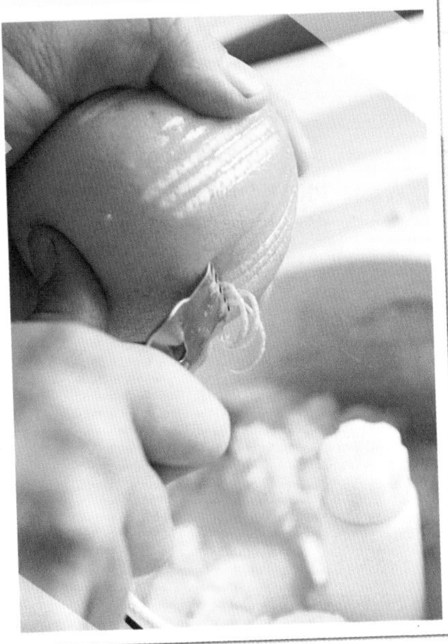

1

STORING FRESH PRODUCE

INTRODUCTION TO

STORING FRESH PRODUCE

Fruit and vegetables are always at their tastiest the moment they have been picked. In fact, it's believed that it is best to harvest a leafy salad crop in the morning, when the sap is rising, and root vegetables in the evening, when the minerals are returning down into the main body of the plant. If you want to keep fresh produce in peak condition for as long as you can, it is very important to set up a good storage system. Here are a few essential rules to keep in mind.

GUIDELINES

- Take the time to trim off any leaves, stems and damaged bits before storing.
- Use a stiff brush to remove most of the soil clinging to root vegetables.
- Don't clean your produce with lots of water to remove all the mud, as it protects the skin; wipe clean with a dry cloth or kitchen paper instead.
- Discard anything that is damaged or diseased – it'll only get worse.
- Try to keep different fruit and vegetables spaced apart. The ethylene that develops from some produce can ripen others more quickly.
- Don't seal produce in plastic – unless you are keeping it in the fridge and using plastic bags specifically designed for food storage.
- Check stored produce regularly.
- Remove any produce that has gone mouldy and think about what you can do to change conditions for the better.

LOCATION

The most important factor is accessibility. The last thing you want on a cold day is to have to go out in the rain and return to the kitchen wet and miserable, so store things inside or keep your storeroom close to the house.

TEMPERATURE

Temperature is extremely important in maintaining a successful storeroom. Most produce that you will store, with the exception of vegetables such as squashes and pumpkins, lasts longer in cool conditions. The ideal temperature range is between 0°C and 4.5°C (32°F and 40°F), so install a thermometer and check it regularly. You will want to draw cold air into your storeroom and keep that cold inside, so it is worth installing an intake pipe low down in the room to draw in cold night air, and another vent at its highest point to let out any warm air. This will also help to encourage air circulation and improve ventilation (see page 21). Remember to screen these pipe openings to keep creatures at bay.

If you can set up your storeroom on the shady side of your house, you will be able to further

Nothing beats produce fresh from the garden

Every tomato has a use, some fresh, some preserved

STORING COMMON FRUIT AND VEGETABLES

PRODUCE	STORAGE METHOD	STORING TIME
Apples	Trays	3–4 months
Beetroot	Clamping	3–4 months
Carrots	Clamping	3–4 months
Garlic	Plaiting	3–4 months
Jerusalem artichokes	Clamping	6 months+
Onions	Plaiting	3–4 months
Pears	Trays	2–3 months
Potatoes	Clamping or sacks	3–6 months
Squashes	Trays	2–4 months

reduce the temperature. Another good idea is to install double doors to stop any warm summer air from getting in when you pay the storeroom a visit. Regular maintenance is vital if you want to run a really effective storeroom: watch the thermometer and open the cold air vents if it looks like the storeroom is getting too warm.

HUMIDITY

High humidity is best for the majority of stored produce. Aim for about 90–95%, as this will stop your food from shrivelling. If you'd like to work out the exact humidity of your storeroom, we recommend that you install a hygrometer, a specialist tool that measures humidity. Simply scattering water on the floor, or placing pans of water on the ground, is the easiest way to increase humidity levels. As a rule, earth floors are better for higher humidity levels than concrete. If your storeroom has an earth floor, cover it with gravel so that it is easy to spread water without getting your feet wet when you go to fetch something for dinner.

Another technique for keeping your vegetables looking fresh is to pack them in damp sawdust or sand. Avoid wrapping produce in plastic bags as this can quickly lead to stagnant air, and surface mould growing on your harvest.

VENTILATION

Movement of air through and within your storeroom helps to remove any ethylene gas, which hastens the ripening process and can make produce smell bad or start sprouting. Leave gaps between your shelves and the wall so that air can circulate freely behind produce. This will also reduce the risk of your fruit or vegetables developing mould.

If you are storing produce on a smaller scale in a cupboard, drill some holes in the door and cover them with fly mesh – anything you can do to improve air-flow is a good thing.

LIGHT

Almost all kinds of fruit and vegetables will store for longer if you keep them in the dark. The exceptions are onions and garlic, which may start to sprout, so it's better to keep them within easy reach in the kitchen.

SHELVING AND TRAYS

Make sure you incorporate plenty of useful storage space if you have a storeroom. We have found that slatted crates make great storage trays because they allow good air circulation and lots of them can be fitted tightly together. Avoid using metal shelving or treated wooden shelves (as there are chemicals in the preservatives) and instead use hard, durable woods such as oak, beech or chestnut. Try to make room for air to flow at the back of the shelves by leaving gaps. You can coat your shelves in plastic so that they are easy to wipe clean at the end of the season.

REFRIGERATION

Using a fridge to the best of its ability sounds like common sense, but it really is the twenty-first-century answer to storing most fresh garden produce. Keep the produce in your fridge in sealed plastic food storage bags, but use paper bags for mushrooms. Keep apples, avocados, melons and other ethylene-emitting produce away from ethylene-sensitive foods such as cucumber, peppers, asparagus and members of the brassica family, as it can result in yellowing, pitting and brown spots. Some foods also emit odours – a small bowl of bicarbonate of soda kept in the fridge will help to absorb any odours and moisture.

METHOD #1

HANGING GARLIC

Garlic is easy to grow and very easy to cook with. We chop it into pesto, butter and dips, use it in roasting trays and even preserve it soaked in honey. Garlic can be used fresh and 'wet', right out of the ground. But if you grow your own, you'll need a storage method that keeps it as fresh as possible. If you allow the outside to air-dry, and hang garlic, either plaited or bundled together, the cloves will retain their freshness for up to 6 months. The same method can be used to store onions and shallots.

PREPARING FRESH GARLIC

Indoors, garlic dries particularly well on simple trays with chicken-wire bases or wooden slats. We made several one year when we had a large garlic harvest, and stacked them on top of each other to save space. In good weather, you can allow the lifted garlic bulbs to lie on the ground outside to dry. However, sometimes it can be too wet and garden space may be at a premium, so you can move them to a corner of a polytunnel or greenhouse to speed up the process. You'll know the garlic is ready for the next stage when the stems are still flexible but yellow-brown and would easily light with a flame.

PLAITING

It's easiest to plait sitting down. Select 3 bulbs and place them on your knees, with the stems pointing away from you. Now plait the 3 together for a couple of centimetres, keeping the bulbs close together and the weave tight. Add another bulb between 2 of the bulbs and repeat the plaiting for another couple of centimetres. Keep going, using your fingers to clamp the plait tight while you make a string with as many bulbs as you like. We usually stop at about a dozen. When you have finished, use a piece of raffia or string to tie the plaited stems together and cut them to an appropriate length. Then tie a knot in the string and hang the garlic on a hook.

Anyone can plait

BUNDLING

A French-style 'grappe' (bunch of grapes) is quicker and easier to make than a plait. All you do is join the bulbs together one at a time, keeping your finger and thumb clamped on the stems and tying each new bulb of garlic into the bundle with string. Stop when you get to about 6–10 bulbs and hang in the same way as a plait.

STORING

Hang your garlic in a light area with good air circulation. Keeping it on a hook by a kitchen window is ideal. Storing garlic in a dark place may accelerate the growing process and encourage it to sprout fresh green shoots. Consume within 6 months.

HOW TO MAKE A GRAPPE

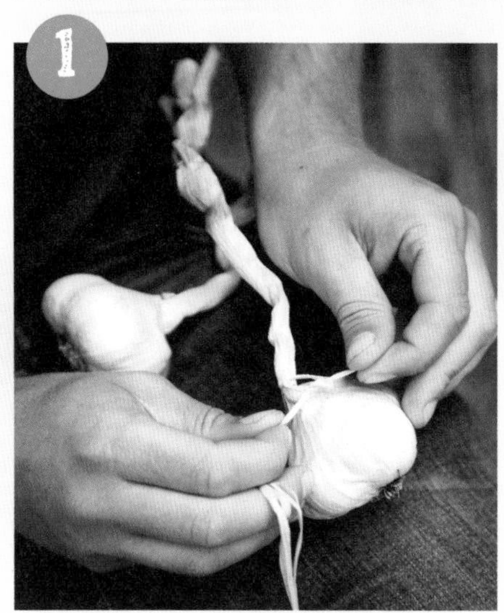

1

Tie a piece of string around the first bulb of garlic, then use the string to make figure-of-eight loop that weaves tightly around the neck of the next bulb, joining the two together.

2

Continue to add bulbs in this way until you have 6–10 bulbs of garlic in your grappe. Fasten off and make a loop that you can use to hang up the grappe.

METHOD #2

CLAMPING ROOT VEGETABLES

Clamping is a traditional storage technique that is very easy to do if you have a bit of space in your garden. It's a fantastic way to keep root vegetables, and we've been clamping produce since we started growing root crops, such as potatoes and parsnips, on a larger scale. Cover your pile of produce with sand, insulate with a layer of straw, then cover loosely with earth, allowing some air to reach the stored food - your crops will keep for between 3 and 6 months.

PREPARING TO CLAMP

Clamped vegetables don't need to be completely clean, but they do need to be in perfect condition. Check each one carefully to make sure it is not damaged. You can give them a quick brush if you like, but the soil will help to preserve your vegetables for longer.

CLAMPING IN OPEN AIR

Harvest your vegetables, leave on top of the soil to dry for a couple of hours, then lay them on top of a layer of straw or bracken. Arrange them into a pyramid shape and cover with more straw. Leave them to sweat for an hour or two, then cover with 15cm (6 inches) of earth. Allow some straw to poke through the soil and enable air to reach the crop. Pat the sides flat with a spade so that they are steep and the rain can run off easily.

CLAMPING IN A BUCKET

Dig a hole in the ground and fill the base with a few stones. Drill several holes in the bottom of a bucket. Spread an 8cm (3 inch) layer of sand or sawdust in the base of the bucket, followed by a single layer of vegetables. Cover with a layer of sand or sawdust and repeat the layers to fill the bucket. Place the bucket in the hole and cover with a 27cm (10–11 inch) layer of straw and then a 5cm (2 inch) layer of earth. This mulch will help to stop the vegetables freezing over the winter. With smaller volumes of vegetables, you can store the bucket with its straw and earth mulch, uncovered, in a cellar or cool shed.

HOW TO CLAMP CARROTS

Put a layer of sand or sawdust in the bottom of your bucket and place a single layer of carrots on top of it.

Repeat these layers until you have filled the bucket, then put the bucket of carrots in a hole lined with some stones.

Cover the bucket with a thick layer of straw.

Now cover the straw with a layer of earth. Mark clearly where the carrots are being stored so that you don't forget.

METHOD #3

STORING SQUASH & APPLES

Keeping produce fresh means that you have more time to use it and also an extended opportunity to decide how you want to preserve it. Squash are famously good for keeping in dry storage, as are apples. Apples that ripen early tend to be eaten straight away, while later fruit is usually good to store. As with all stored and preserved produce, you will need to keep an eye on them.

SETTING UP A STOREROOM

All you need to store fresh produce is a room with good air circulation and some shelves and trays. The amount of space you need will depend on how much you grow. To store, simply spread your produce out on a tray with a little separation between each item, then place the tray on a shelf. Most shelving has enough of a gap between shelves for general household storage, but you might need to customize your shelving to meet your needs. Apples, for example, need only 10cm (4 inches) between shelves, while the amount of room needed for squash will depend on the variety you grow.

An easy way to check if it is ready

Get into the habit of visiting your storeroom regularly. This will stop you buying food you already have in stock and give you the opportunity to check that nothing is spoiling.

STORING SQUASH

A squash or pumpkin is ready to store when the skin is hard enough to withstand your fingernail being pressed into it, and when the stem connecting the fruit to the plant is shrivelled up and dry. Be selective and avoid damaged ones. You can always make a few soups with any squash that are not suitable for keeping and freeze them.

Place the perfect specimens on trays lined with newspaper or straw, and keep in a cool, dark place. Check regularly and discard any that show signs of collapse. If conditions are right, your harvest will keep well into the spring. (We rarely have any left by then as they are too good to have sitting around.)

STORING APPLES

The old saying that it takes only one rotten apple to spoil the barrel is sadly true. Any apples that are visibly bruised or damaged will go off and risk spoiling the rest of your crop. Therefore, select only the best-quality apples for keeping, and eat or cook with the rest.

Wrap the keepers individually in brown paper or newspaper. Try to use the palms of your hands instead of your fingertips to avoid damaging the fruit – you will be surprised how delicate some varieties are.

Place the wrapped apples either in apple trays or on shelves, making sure they do not touch each other to allow air to circulate and reduce the risk of spoiling.

SPICED APPLE SAUCE

This is a fantastic way to use your fresh apples. Delicious served on its own or with either roast pork or pheasant.

Makes 2.5kg (5½lb)

- 25g (1oz) butter
- 1kg (2lb) apples, peeled, cored and chopped
- 1.25kg (2½lb) sugar
- 500ml (17fl oz) cider
- a pinch each of cinnamon, cloves and allspice

Melt the butter in a large pan, then add the apples and cook gently until softened. After a few minutes add the sugar and stir, then add the cider and spices. Bring to the boil, then reduce the heat to medium and simmer for 15 minutes. Strain through a sieve to remove excess liquid and store in a sterilized jar (see page 53) in the fridge.

DRYING

INTRODUCTION TO

DRYING

Drying is one of the oldest ways to preserve food. If done properly, using gentle heat, drying does not destroy as many of the nutrients as some of the other preserving techniques that require heat, such as canning and bottling. In warm climates, if food is harvested, spaced so that dry air can circulate around it and then left for a period of time, the air will take the moisture out of it. In cooler and more humid climates, however, it is important to monitor the conditions, and, if necessary, assist the drying process.

HOW DRYING WORKS

The variety of produce that benefits from drying is impressive: you can dry cured meat, vegetables, herbs, spices, fruits and fungi. Taking the water out of produce deprives micro-organisms (bacteria, yeast and mould) of the moisture they need in order to survive and spoil the food. It also prevents chemical reactions (enzymatic deterioration) from destroying the food during storage. Although drying itself does not destroy enzymes, most dried food is low enough in moisture to prevent enzymatic deterioration. If you can remove 80–90% of the moisture, the microbes are unable to perform.

Drying may prevent microbial growth, but bear in mind that there are chemical reactions caused by enzymes that can occur unless the food is pre-treated before drying. Thankfully the enzymes in vegetables can be counteracted by simply blanching them (a quick immersion in boiling water or steam), and you can counteract the enzymes in fruit by treating the food with ascorbic acid, a form of vitamin C.

Dry at the lowest temperatures, otherwise the food will cook rather than dry. If the temperature is too high the cooking will cause the outside of the food to go hard and stop the moisture escaping. Turning up the heat to reduce the drying time is therefore counter-productive. To speed up drying, a low humidity is essential, and increasing the air-flow over the food surface is also beneficial.

DRYING METHODS

There are numerous ways of drying food: using the sun, leaving the food out in the fresh air for a period of time, or using a commercial dehydrator. We have hung things up to dry in porches, used our fan oven with and without heat, and have built solar dryers (see pages 38–9), though they are only really effective in the middle of summer. Generally, air drying is complete in 2 or 3 days, depending on conditions (see pages 34–5).

FRUIT DRYING

Ripe fruits are best, but try not to use any that are too ripe or over-ripe, as the results will not be as high in quality. There are no hard and fast rules about how to slice fruit for drying. Some fruits (such as figs and berries) are best dried whole, but generally thin, even, peeled pieces dry the fastest. You can

Colourful and healthy - dried root vegetables are full of flavour and goodness

DRYING COMMON FRUIT AND VEGETABLES

FOOD	PREPARATION	TREATMENT	DRYING TIME
Apples	Peel, core and cut into rings or slices	Dip in lemon or orange juice for 4 minutes	4–6 days
Figs	None	None	4–5 days
Grapes	Choose seedless varieties	Blanch for 30 seconds	3–5 days
Aubergines	Cut into 5mm (¼ inch) slices	Blanch for 4 minutes	6–8 days
Green beans	Wash and cut into short lengths	Blanch for 2 minutes	8–10 days
Mushrooms	Brush any compost off the mushrooms and slice large ones	None	6–8 days
Onions	Slice about 3mm (⅛ inch) thick	None	8–11 days
Peppers	Cut into 3mm (⅛ inch) rings	None	6–8 days
Parsley	Wash and cut along stalks lengthways	None	6–8 days

leave the peel on, but the process will take a little longer. Prior to drying, some sliced fruit should be dipped in orange or lemon juice for a few minutes (see table above). As there are sugars in fruit, it can become sticky as it dries, so use a non-stick baking sheet, or cover your baking sheet with greaseproof paper. Place items in a single layer and avoid overlapping.

VEGETABLE DRYING

Onions, peppers, celery and mushrooms are all well suited to drying. First cut the vegetables into pieces of uniform size and thickness. If they need to be blanched (see table above), immerse the vegetables in boiling water for a couple of minutes, then strain through a sieve and immediately immerse in iced water to stop the cooking process. Drain, then dry.

HERB DRYING

The best time to harvest most herbs for drying is just before the flowers open, when the buds are about to burst. Drying can take 2 or 3 weeks, depending on the conditions. Rinse the herbs in cold water, then shake off any

excess. Those with long stems can be dried in bunches. Tie the stems together and hang them upside down where there is plenty of room for air to circulate. If the environment you intend to hang them in is dusty, you can place each bunch in a paper bag with some holes cut in it – the air should still circulate, but the dust will not settle on the herbs. Large-leaved herbs such as basil, or those with short stems, can be dried on a rack or tray. If you use a rack, turn the herbs over daily.

FOOD DEHYDRATORS

Food dehydrators are small electrical appliances for drying foods indoors, and have a heater and a fan. They are efficient and reliable, but make sure you get one that is big enough for your needs. Remember that your oven can also be used as a food dehydrator. If the oven is fan-assisted, it is even better.

DRYNESS TESTING

Make sure your food is thoroughly dried before storing. The best way to determine the dryness of the food is to consider its look, feel and taste. Take several pieces of food and cut through the centre of the thickest part. There should be no visible signs of moisture (a dark centre means the food needs to dry for longer). Check the following:

- Fruit should be pliable and springy, and should not stick together if folded over on itself.
- Vegetables should be brittle, as if they would shatter if hit with a hammer.
- Meats should feel very dry, and jerky, for example, should be dark, fibrous and form sharp points when bent.
- Herbs should be brittle and should crumble if rubbed in the palms of the hands.
- Fruit leathers should be dry to the touch and peel away easily from the tray.

STORING

Having achieved the dryness you are after, it's important to store what you have made. Sterilized sealable jars are the perfect way to store your dried produce (see page 53).

METHOD #4

AIR DRYING

Air drying is one of the simplest and most decorative methods of preserving. Not all fruit and vegetables are suited to it, but there are some that will transform your kitchen into a hanging garden of edible decorations. Every year we especially like to hang up chillies and herbs ready for use over the winter. By removing moisture there is far less chance of your produce spoiling before you get the chance to use it, and the drying process can also intensify the flavour, which will be released when the dried ingredient is added to your cooking.

DRYING CHILLIES

Use string to tie the chillies into clusters of 3 or 4 at regular intervals, then suspend the clusters in a line, using hooks. Alternatively, pass a needle and thread through the stems. This way you'll end up with a long, even line of chillies with plenty of room for air to circulate.

Start by choosing a good place to hang your chillies. An airing cupboard could dry them in a few days, and somewhere in the kitchen (ideally not in a corner as that can hinder air circulation) can air-dry herbs or chillies in 2–3 weeks. If you have a porch or lean-to, try hanging a string of chillies under the eaves. The key is for the chillies to remain dry but have good air circulation around them. We've learnt from experience that although drying above the oven looks great, and they can dry more quickly thanks to the extra heat, it's not worth doing because the chillies get covered in a thin, oily residue from the pans cooking below.

Leave for 2–3 weeks, or until they are dry. Check them every couple of days. When the chillies are so dry that the pods snap instead of bending, they can be stored in glass jars. Keep in a dark, cool place.

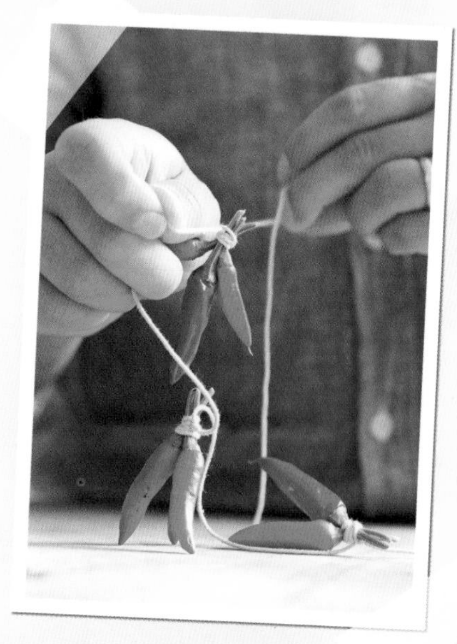

Any knot will do the job

LAZY CHILLI FLAKES

150ml (5fl oz) vinegar
2 tablespoons sugar
50g (2oz) dried chillies, stems removed
a pinch of salt
a pinch of freshly ground black pepper

Put the vinegar and sugar into a pan and heat gently until the sugar has dissolved. Blitz the chillies in a food processor, or grind them using a pestle and mortar, until flaky, and add to the vinegar. Bring to the boil, then allow to cool in the pan. Season with salt and pepper, then pour into a jam jar, seal tightly and store in the fridge.

There's a lot of heat in this jar

DRYING HERBS

For the most intense flavour, pick herbs such as sage, rosemary and thyme just before they flower, wrap their stems tightly with string and hang them in small bunches somewhere airy. Once dry, take them down and pluck the leaves from the stems. Put the aromatic leaves into jars and keep close to the cooker for ready-to-use flavour impact.

If you don't grow your own herbs, ask neighbours or friends whether you could harvest some herbs from their garden, as most herbs will benefit from the occasional trim. You could buy bunches of herbs for drying, but that's not an economical option.

Dried chillies are integral to much of the cooking that we enjoy. This recipe combines dried chilli flakes with herbs and makes a very tasty barbecue kebab. Using the entire chilli, seeds and all, gives this dish plenty of heat.

SERVES 4–8

400g (13oz) monkfish tail, cubed
400g (13oz) salmon fillet, cubed
100g (3½ oz) large tiger prawns
juice of 1 lime

FOR THE CHILLI AND HERB RUB
4 dried chillies, finely chopped
2 garlic cloves, finely chopped
1 tablespoon chopped fresh lemon thyme
1 tablespoon chopped fresh parsley
zest of 1 lime
1 teaspoon sea salt
1 teaspoon freshly ground black pepper
2 tablespoons olive oil

FOR THE SWEET CHILLI SAUCE
2 dried red chillies, finely chopped
1 garlic clove, crushed
2 tablespoons sugar
60ml (2½ fl oz) water
20ml (½ fl oz) white wine vinegar
1 teaspoon cornflour

SEAFOOD KEBABS WITH CHILLI & HERB RUB

Put 8 wooden skewers to soak in cold water for at least 30 minutes.

To make the chilli and herb rub, place the chillies, garlic, thyme and parsley in a large bowl with the lime zest, salt and pepper. Add the olive oil and stir well.

Add the fish and prawns to the bowl, stirring thoroughly to make sure everything is completely coated with the spicy rub. Heat your barbecue or grill, and while it is heating up, thread the chilli fish and prawns on to skewers.

To make the sweet chilli sauce, put the chillies and garlic into a small pan. Add the sugar, vinegar and 60ml (2½fl oz) of water and bring to the boil. Reduce the heat and simmer for 3–4 minutes. Finally, mix the cornflour to a paste with a tablespoon of water and stir into the sauce until incorporated.

Cook the kebabs over or under a high heat for 5 minutes – watch them carefully to avoid burning. Serve with the sweet chilli sauce and a squeeze of fresh lime juice.

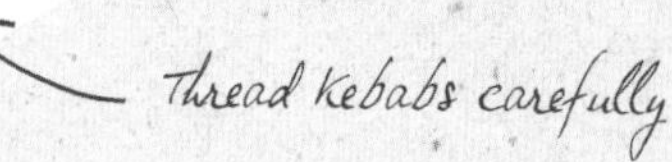

Don't overcrowd the tray

METHOD #5

SOLAR DRYING

Solar drying is a preserving method that's been around for thousands of years and it has begun to make a comeback. At its simplest, solar drying can be done with virtually no equipment – all you need is air-flow around your chosen fruit or vegetable, and then it's a matter of letting nature take its course. We built our own solar dryer to improve our ability to store ripe garden produce, such as tomatoes, chillies and apples, and to save money on luxury sun-dried foods. We're big fans of this method of drying produce – no fuel is required, so the dryer costs nothing to run.

ADVANTAGES AND DISADVANTAGES

- Solar dryers are really easy to build in a couple of hours from cheap materials.
- They are great for summer preserving as they don't contribute unwanted heat inside the house.
- Solar drying is completely free, using no fuel other than the power of the sun.
- Solar dryers take longer than a conventional oven to dry food, so preparation has to be done several hours – or even days – before you intend to eat your produce. But there is no real supervisory cooking time, so we enjoy the time off.
- Cloudy and rainy weather seriously impairs performance, so solar dryers are only really a sunny weather option.

HOW A SOLAR DRYER WORKS

A solar dryer consists of a solar collector, a glass-covered box lined with black, heat-absorbing material (usually a piece of metal, as it's particularly effective at absorbing heat). The solar dryer is angled towards another box – the solar oven – which contains a metal rack, and the heat collected filters upwards to dry the food placed on the rack. The inside walls of the solar oven are painted white so that heat loss is minimized and the warmth is retained in the drying area. The performance of a solar dryer is much slower than that of a conventional gas or electric oven, but as an inexpensive way to preserve produce it is unbeatable.

MAKING A SOLAR DRYER

The first stage is to construct your solar collector, which is essentially a large wooden box with a glass front. The wood helps to insulate the box and keep in the heat. To save time, use an old glass-fronted cupboard to build your solar dryer, or reuse an old wooden box or drawer with an old window fitted over it. The bigger the better for this part of the solar dryer, as you will be able to collect more sunshine and therefore dry food more quickly as a result. Remove one end of the box. Then drill holes in the other end and

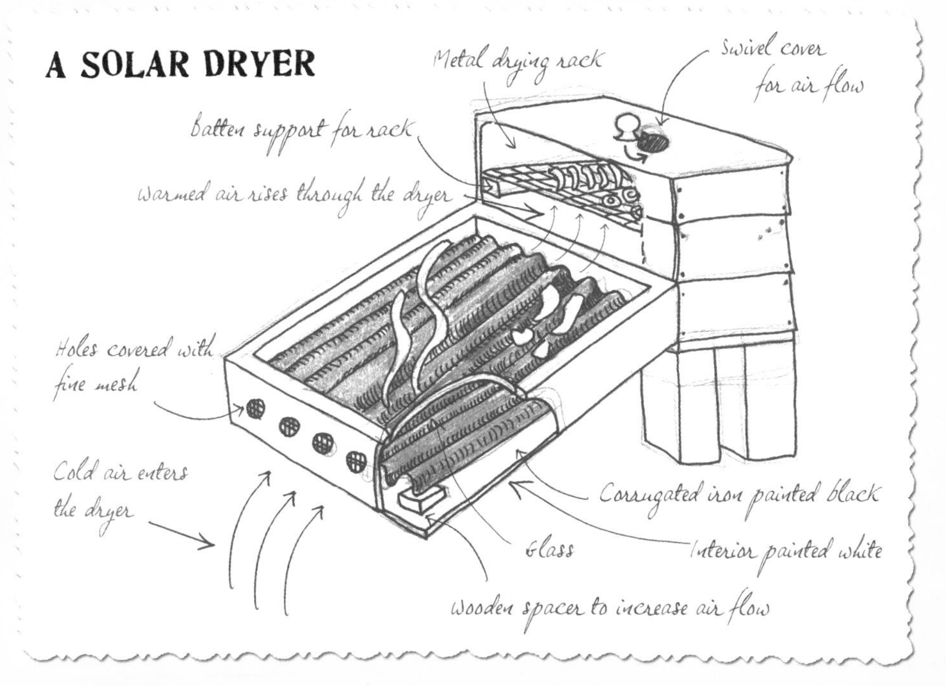

cover them with fine mesh to prevent insects getting into the system. These holes allow air to be drawn in at the base, and this is warmed inside the collector before rising into the oven through the open end that will eventually sit on top. Close up any other holes and gaps. Paint the inside of the box white, and tack on a small blocks of wood to act as spacers for air to circulate around the black material to be fitted next.

Cut a piece of metal, preferably corrugated iron, to fit into the box and paint it black. Place this piece of black material inside the box, sitting it on the spacers. Cover the box with a hinged glass panel or old window. Position the solar collector so that it is directly facing the midday sun and build a support for it to keep it at the correct height and angle. This structure will also support the oven or drying box.

The next step is to build a very simple box to fit above the solar collector. This will be the oven where you put food to dry. Paint the inside of your oven white so that heat doesn't radiate outside when it heats up. Position battens on either side of the oven for placing wire trays or grills on. Drill several holes in the lower front edge of the oven so that they line up with the open end of your solar collector. Seal the two sections together with screws, making sure that there aren't any gaps. Drill a vent hole at the top of the box. Then make a swivel cover by attaching a small door knob to a piece of wood big enough to cover the vent and screwing one corner to the box.

POSITIONING IT

Food in a solar dryer doesn't reach very high temperatures, so if you want to dry something and eat it the same day, it's best to start drying before midday and continue all day. Place your solar dryer in direct sunlight facing south. You'll need to move it every couple of hours to track the direction of the sun and avoid shadows from surrounding objects, but if you don't have time for this, simply position it facing where the sun will be at its highest point of the day. Drying times depend hugely on the specific shape, size and design of your solar dryer, and, of course, on the weather. The hours before and after midday tend to be the most effective for drying food.

USING THE SOLAR DRYER

We recommend using a solar dryer for drying fruit, vegetables, seeds and herbs. Complex dishes can be cooked in a solar dryer, but you really need a very hot sunny day to achieve good results.

Before drying, remember to make sure you cut the food into smaller pieces than normal as a larger surface area yields better results. If the cut fruit or vegetable would normally discolour when left in the air, it will also do that in the solar dryer. To prevent this, rub the sliced food with lemon juice or ascorbic acid before you put it in the dryer.

Don't be tempted to turn the food or peek at what is cooking inside your dryer as

HOW TO USE A SOLAR DRYER

1 Spread out a layer of the food on your grill and place it in the dryer – make sure the grill sits well off the oven floor as the air needs to circulate all around the food.

2 Put the lid on the dryer and leave until the food has dried.

this will allow precious heat to escape and slow down the whole process. The usual temperatures achieved by our solar dryer are around 50°C (120°F), which is ideal for drying herbs.

OTHER DESIGNS

There are many different designs of solar dryer. Another of our favourites is based on the idea of reflecting the sun's rays on to a focused spot in order to achieve higher temperatures. Take an old satellite dish and cover it in tin foil. Face it towards the sun and hang a kettle above it to boil water for a cup of tea!

SUN-DRIED TOMATOES

Thinly slice your tomatoes. Lay these on a wire tray, put them into the solar dryer and dry them in the midday sun for 4–5 hours. When ready to eat, they should feel completely dry and chewy. If they haven't reached that point, remove them from the solar dryer, place them in an airtight container overnight, and repeat the drying process the next day. Depending on the sun, it could take a couple of days or even up to a week to dry the tomatoes.When finished, place them in a jar with olive oil or seal in an airtight container, and place in a dark cupboard. They can be stored for up to 6 months. Rehydrate in warm water for 1 hour before using, or use straight from the oil.

SUN-BLUSHED TOMATOES

Sun-blushed tomatoes are simply partially dried. Follow the instructions for sun-dried tomatoes but remove them from the dryer after the first 4–5-hour drying session. Store them in the fridge with a drizzle of oil and consume within a week.

Sun-blushed tomatoes with a little salt

We love this granola – it's like a breakfast cereal that has been crossed with a flapjack. The fruit and nut combinations are completely yours to decide on – try hazelnuts, Brazil nuts, pumpkin seeds or sunflower seeds. The ratio of roughly equal parts cereal to fruit and nuts keeps a good balance, but otherwise it is all down to personal choice and whatever you've stored in your larder.

MAKES ABOUT 1.25kg (2½lb)

150ml (5fl oz) rapeseed oil

100ml (3½ fl oz) hot water

200ml (7fl oz) honey

500g (1lb) rolled oats

300g (10½ oz) nuts (sliced almonds, chopped pecans and chopped walnuts)

200g (7oz) dried fruit (raisins, sultanas and dried cherries)

LUXURY GRANOLA

Preheat the oven to 160°C (325°F), Gas Mark 3.

Place the oil, hot water and honey in a bowl and whisk to combine. Put the oats and nuts into a large bowl and mix together. Pour in the honey mixture and stir until the oats and nuts are evenly coated.

Spread the mixture on to 2 baking sheets. Bake in the oven for 10 minutes, then remove, stir, put back into the oven and bake for another 10 minutes, or until toasted.

The mixture will harden as it cools. Once the muesli has cooled completely, break apart any large lumps, stir in the dried fruit and store in an airtight container at room temperature for up to 2 months. Alternatively, you can freeze the granola for up to 6 months.

METHOD #6

OVEN DRYING

Oven drying food is a really effective way to preserve produce beyond its natural growing season and effectively prevents the development of unwanted enzymes, bacteria, yeasts and fungi that thrive on moisture. Not only does dried produce last longer, it often has a more intense flavour, and you can eat sweet summer fruits and autumnal treats in the heart of winter. On the financial front, drying your own fruit and vegetables is a fantastic way to enjoy luxury ingredients such as dried mushrooms, chillies and sun-dried tomatoes at a fraction of the cost - especially if you use the residual heat in the oven after cooking a meal or baking a cake.

CHOOSING YOUR PRODUCE

We only ever dry the freshest, healthiest and most prolific of our fruit and vegetables, and make sure we use sharp tools when harvesting to avoid bruising the produce. Try to time it so that you pick your plants when they are not covered in dew and moisture, as this will slow down the drying time – wait until the afternoon for best results.

DRYING

To dry fruit and vegetables, you have to slice them, spread them on trays and heat them at a low temperature for a long time. Removing the moisture slowly in a conventional electric oven for a few hours at about 45–55°C (113–130°F) works well. If your oven doesn't go down to 50°C, set it to its lowest temperature and leave the door open. Dry several things at once to make full use of the heat.

IS IT READY YET?

Testing to see when your produce is dried is a simple matter of trial and error. Dry your chosen food for a few hours, then test to see if it needs longer. If it does, leave it in the oven for another hour and test again. Repeat until the food is dry. Try to close the door quickly after looking in order to save the stored heat.

With plants such as herbs or chillies, test them by bending them in your hand – if they snap, they're ready, if they are bendy, leave them for a little longer. With some produce, including most fruits, you will know it is ready if you squeeze it and no juice comes out.

STORING

Storage containers are a key part of the drying process. Airtight containers, food-grade plastic bags and glass jars are all good ways to store your dried food. We save almost every glass jar that comes into the house, cleaning and reusing them for storing our dried produce. If you are planning on drying in bulk, large Kilner jars are a good investment.

NOW TRY: APPLE & PEAR RINGS

Core the fruit, cut it into rings 5mm/¼ inch thick, then dip them into a solution of water, lemon juice, a teaspoon of sugar and a pinch of ground cinnamon. Blot off any excess liquid with kitchen paper and place your rings on a rack in the oven at about 45–55°C (113–130°F) for 3–5 hours. When they are ready they can be eaten dry as a snack or used in cooking. Store for 3–6 months.

NOW TRY: ROOT CRISPS

Beetroots, parsnips and carrots all make excellent crisps, and luckily this healthy snack is easy to make. Clean the vegetables, then slice thinly, place on a baking tray and put them in the oven at about 45–55°C (113–130°F) for 3–5 hours. When they are ready, place in an airtight container and store for 3–6 months.

NOW TRY: HERBS & SPICES

Dry leafy herbs, such as rosemary and sage, just before they flower. This is when their oils are at the highest level and the leaves are highly aromatic. Arrange your herbs on a baking tray and dry in the oven at 40°C (105°F) for 2–3 hours. Once dried, hold them over a piece of paper and crumble between your fingers to remove the leaves from the stems. Tip the herbs into an airtight container and store in a dark place for up to 6 months. The same process can be used for drying the seeds of herbs such as coriander and cumin. A maximum temperature of 47°C (115°F) is the ideal, as the aromatic oils can be damaged by higher temperatures.

strip the leaves off the stem

Home-grown strawberries are abundant in the summer months but unavailable for the rest of the year, so we like to make them into a snack that will last a little longer than the fresh fruit. The secret is slow cooking at a very low temperature. This recipe works well with any soft fruit, so you can experiment with other ingredients too.

SERVES 4

300g (10oz) strawberries

50g (2oz) sugar

STRAWBERRY LEATHER FRUIT SNACK

Preheat your oven to 50°C (122°F), or its lowest temperature.

Now put the strawberries in a bowl and crush or blend them until they're fairly smooth but with a bit of texture remaining. Add the sugar to the bowl and stir well.

Spread the fruit mixture on to a piece of baking paper or a non-stick baking tray and place in the oven for 6–8 hours.

Let cool on the baking paper or tray, then cut into long, thin strips and peel each strip off individually. Roll them up and store in an airtight container for up to 3 months.

3

JARS

INTRODUCTION TO

JARS

A glass jar with a sealed lid is the perfect container in which to keep food for long periods of time. There is something very satisfying about going to the larder and choosing from a plethora of colourful jars, all of which you have lovingly filled. When you take control of your larder and know exactly what has gone into each recipe, you can rustle up a meal that represents the very best of home cooking. There are a number of ways to prepare food to be stored in jars: in this section we cover preserving in brine, vinegar, oil, sugar and alcohol. Some produce lends itself to a specific type of treatment, but each method produces its own distinct flavours and textures.

KEY METHODS

The method we focus on here involves storing home-processed food in sterilized glass jars. Food that should be preserved in this way has high acidity (a pH of more than 4.6), and high salt or sugar content, for example jams, pickles and chutneys, as well as most fruits.

For the record, there is also a method whereby home-processed food is preserved by sealing it in mason jars. This is called 'canning', and the food can have a shelf-life of 1–5 years. Food that should be preserved in this way has low acidity (a pH of less than 4.6), and includes most vegetables as well as meat and poultry. If not done correctly, home canning can expose you to botulism (a serious form of food poisoning), so the food must be brought to a high enough temperature (116–130°C/240–265°F) for long enough to kill the micro-organisms that carry disease. As you don't want to the food to become unpalatable and lose all nutritional value, you should aim to get a temperature of 121°C (250°F) for 3 minutes, using a large, robust pressure cooker designed for the job.

CHOOSING YOUR JARS

Numerous delicatessens tempt us with gorgeous jars filled with artisan-made jams, chutneys, pickles, mustards ands sauces. If you agree that there is something rather decadent about paying over the odds for such prettily packaged produce, you will immediately understand how your own jars filled with delicious products can make excellent presents – the perfect combination of luxury and thoughtfulness. Always keep your eyes out for attractive jars that will make unique gifts.

Before you launch yourself into filling every available jar you can find, it's important to understand why certain things are essential. First, the food must be treated or processed so that the micro-organisms that can cause it to spoil are unable to function. Second, the jars must be sterilized to ensure that the jams, pickles and preserves you spend a lot of time making will keep well. Jars that have not been sterilized properly will infect the food inside so that it will spoil very quickly and have to be thrown away.

Herb-infused oils look good and taste great

Every bit of juice goes into a marmalade

HOW TO STERILIZE JARS

There are numerous ways of sterilizing jars, but we find the following method to be extremely effective.

- Heat the oven to 140°C (275°F), Gas Mark 1.
- Arrange the jars and lids on the oven shelf, making sure they are not touching each other.
- Always prepare more jars (and lids) than you think you will need. Should you end up with more mixture than anticipated, it is too late to start sterilizing more jars once the food is ready.
- Check the lids to ensure they are free from nicks and cracks, otherwise they may not be airtight.
- Close the oven door and leave the jars to heat up for about 20 minutes.
- If using a funnel to fill your jars, be sure to sterilize it too (if the funnel is made of plastic, sterilize it by immersing it in boiling water for a few minutes).

HOW TO FILL JARS

- Using thick oven mitts, remove each jar from the oven as needed and put it on to a heatproof mat or heat pad.
- Make sure you fill the jar while the preserve is as hot as the glass.
- Do not add cold food to hot jars, or hot food to cold jars, otherwise the jars might crack.
- Leave a little 'headspace' at the top of each jar before you seal it – it's necessary to assure a secure vacuum seal.

HOW TO SEAL JARS

- Always wipe the rim of the jar before you seal it, because any food debris or traces of salt left behind will prevent it from sealing properly.
- Leave jams, preserves or pickles to settle for 15 minutes with the lids just sitting on the top before sealing.
- Make sure to label and date your jars – it will make managing your store cupboard much easier in the long run.

HOW TO CHECK JAR SEALS

When hot food is put into a sterilized jar and the lid is screwed down (or, if you are using clip-top Kilner jars similar to the ones illustrated below, when the hinged lid is snapped into place over the sealing ring), the air in the gap above the content will contract as it cools, creating an airtight seal. Lots of modern jar lids have dimples that are pulled down when this happens, and if you press on the lid, there should be no popping noise. If it does move and 'pop', the contents may not be safe to eat. The lid of a securely sealed jar will always be firmly sucked down on to the top of the jar.

Cold pickles keep their crunch

METHOD #7

PICKLING

Cold pickling preserves the texture and appearance of your ingredients, so cold-pickled cucumbers, for example, retain their vibrant colour and crispiness. The preparation involved in making cold pickles is largely to do with getting the vegetables ready – peeling, chopping and slicing them as necessary. The basic technique requires the use of salt to draw out as much moisture as possible from the vegetables before you pump them full of the preserving power of vinegar and the flavour of pickling spices.

HOW TO MAKE PICKLED GHERKINS

Slice your cucumbers into thin strips and layer them in a bowl or plastic container. Cover lightly with salt, put some garlic cloves on top, followed by more salt. Repeat the layers and leave overnight.

Rinse thoroughly, then place in sterilized jars (see page 53). Mix the shallots, vinegar, bay leaf, herbs and mustard seeds in a jug, then pour over the cucumbers, making sure to fill the jar. Seal the jar, leave the pickles to infuse with flavour and enjoy 3–4 weeks later.

PREPARING FLAVOURINGS

If you want to save time when pickling, it is worth making a batch of pickling spices and storing them in their own clearly labelled jar. A classic pickling spice recipe would include a variety of broken and crumbled herbs and spices, the more fragrant the better. Some of our favourite flavourings are mustard seeds, black peppercorns, cloves, allspice, mace, bay, ginger, coriander seeds and bashed cinnamon sticks.

MAKING YOUR PICKLE

Start by preparing your vegetables. Cut off any stalks or roots, then carefully peel, chop and slice the vegetables as required. Start layering them in a bowl or plastic container with thin layers of salt between. Repeat this process until all the vegetables that you want to pickle are covered in salt. Leave overnight at room temperature.

The next morning, wash off the excess salt under running water. Pack the vegetables into sterilized jars (see page 53), mix your vinegar and flavourings together, then pour the mixture over the vegetables in the jars, making sure they are fully submerged.

Seal the jars and store for at least a few weeks before enjoying. Cold pickles will keep for 6 months.

PICKLED GHERKINS

Makes 500g (1lb)

- 500g (1lb) knobbly cucumbers, finely sliced into long thin strips
- 200g (7oz) salt
- 2 garlic bulbs, cloves peeled
- 3 shallots, finely sliced
- 800ml (24fl oz) white wine vinegar
- 1 bay leaf, crumbled
- 2 tablespoons chopped fresh dill or tarragon
- 2 teaspoons mustard seeds
- vine leaves (optional)

Follow the instructions opposite to make pickled gherkins.

PICKLED ONIONS

Makes 1kg (2lb)

- 1kg (2lb) onions, peeled
- 100g (3½ oz) salt
- 200g (7oz) sugar
- 1 litre (1¾ pints) malt vinegar
- 1 teaspoon coriander seeds
- 1 teaspoon yellow mustard seeds
- 1 teaspoon allspice
- 1 teaspoon black peppercorns
- 2 dried chillies

Sprinkle the onions with salt and leave overnight. Rinse and put them into sterilized jars (see page 53). Place the sugar and vinegar in a pan and heat until the sugar has dissolved. Add the pickling spices, then pour over the onions. For best results, leave for a month before eating.

Hot and tangy

METHOD #8

HOT PICKLING

As the title suggests, hot pickling involves applying heat to at least some of the ingredients. A good example of this technique is lime pickle, which, like many pickles made from fruit, is intended to alleviate the heat of curries. There are any number of variations, but it's all up to personal taste. The texture is determined by the size of the pieces of citrus fruit. Thin slices or smaller chunks will soften more during the cooking process (a lime cut into 16 pieces will be thinner and softer than one cut into 8). Not surprisingly, the hotness of the pickle can be changed by simply varying the amount of chilli used: 5 tablespoons makes it nice and tangy, but 10 or more and the pickle becomes seriously hot.

HOT LEMON PICKLE

Lemons can be used instead of limes, but they break down more quickly so the texture is softer.

- 1kg (2lb) lemons
- 100g (3½ oz) salt
- 3 tablespoons mustard seeds
- 10 fenugreek seeds
- 2 teaspoons cumin seeds
- 1 teaspoon cardamom seeds
- 400g (14oz) sugar
- 5 tablespoons chilli powder
- 3 garlic cloves
- ½ teaspoon asafoetida

The method is the same as for lime pickle (just follow the instructions opposite).

HOT LIME PICKLE

- 1kg (2lb) limes (about 16)
- 100g (3½ oz) salt
- 3 tablespoons mustard seeds
- 10 fenugreek seeds
- 2 teaspoons cumin seeds
- 1 teaspoon cardamom seeds
- 400g (14oz) sugar
- 5 tablespoons chilli powder
- 75g (3oz) fresh ginger, grated

To make the lime pickle, follow the instructions opposite.

HOW TO MAKE HOT LIME PICKLE

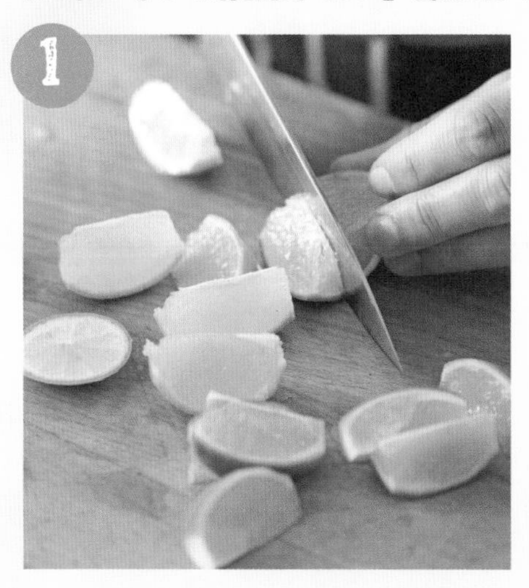

Soak the limes overnight in a bowl of cold water. Drain, then cut the top and bottom off each lime. Chop the fruit into pieces, removing the pips to avoid bitterness.

Place the limes in a large bowl and sprinkle with the salt. Leave for several hours so that the moisture is drawn out of the limes.

Mix the aromatic seeds together. Pour the liquid from the bowl of limes into a pan, and add the seeds and sugar. Bring slowly to the boil, stirring until the sugar dissolves.

Boil for a minute, then remove from the heat and add the chilli. When cool, stir in the limes and ginger. Pour into sterilized jars (see page 53). Keep in a warm place for 4 or 5 days, then store in a cool dark place for 4 weeks.

The velvety sauce and crisp vegetables are what make a good piccalilli so special, and this recipe has both. We love it with cheese, cold meats and sausages – the list is endless.

MAKES 2.5–3kg (5½–6½lb)

1 cauliflower, broken into small florets

1 courgette, cut into 1cm (½ inch) dice

3 medium onions, cut into 1cm (½ inch) dice

3 carrots, cut into 1cm (½ inch) dice

1 cucumber, peeled, seeded and cut into 1cm (½ inch) dice

200g (7oz) green beans, cut into 1cm (½ inch) pieces

100g (3½ oz) salt

75g (3oz) plain flour

1 tablespoon mild curry powder

15g (½ oz) ground turmeric

45g (1¾ oz) mustard powder

25g (1oz) ground ginger

1 litre (1¾ pints) white (distilled) malt vinegar

600g (1lb 3oz) sugar

PICCALILLI

Put all the prepared vegetables into a large bowl, toss with the salt, then cover and leave for 24 hours. Rinse well in several changes of water and leave them to drain in a colander for 30 minutes.

Sift the flour and spices into a large pan and gradually start to add the vinegar, blending well until you have a smooth paste with no lumps. Add the rest of the vinegar and the sugar and bring to the boil, stirring all the time until thick. Add the prepared vegetables and bring back to the boil for 2 minutes.

Remove the pan from the heat, pour the piccalilli into sterilized jars (see page 53) and seal. Leave for at least a month to mature before eating.

The jars should keep, unopened, for 18 months if stored in a cool dark place. Once opened, keep in the fridge for up to 6 weeks.

Fresh and crispy vegetables

Sometimes we all need a satisfying bite in a hurry. This recipe for sauerkraut is much quicker than the traditional fermentation method. We make a batch of it every so often to eat in a sausage sandwich, or to serve with roast pork.

SERVES 4

- 1 tablespoon olive oil
- 1 onion, thinly sliced
- 1 white cabbage, shredded
- 300ml (½ pint) cider vinegar
- 125ml (4fl oz) cider
- 1 tablespoon salt

QUICK SAUERKRAUT

Heat the oil in a heavy-bottomed pan, then add the onion and cook until it softens. Add the cabbage, vinegar, cider and salt and bring to the boil. Simmer for about 30 minutes, adding water if it gets too dry.

Stored in a sterilized jar (see page 53), this simple sauerkraut will keep for 2 weeks in the fridge.

Pour in the cider vinegar

Any bottle will do

METHOD #9

FLAVOURED VINEGARS

Vinegar is often a tool rather than a star in the preserving process. Perhaps that is because the usual ones – cider vinegar, white and red wine vinegar and malt vinegar – don't often have any distinctive flavour to speak of. We have found that there is room to be incredibly experimental with vinegar, when making colourful concoctions and seasonal preserves. Why not give it a go and become a vinegar chef? The flavoured vinegars on these pages will keep for up to a year.

HOW TO MAKE RASPBERRY VINEGAR

1. Combine the vinegar, sugar and raspberries in a pan and simmer for 15 minutes on a medium heat.

2. Strain the mixture through a fine sieve and pour into a sterilized bottle.

HERB VINEGARS

Start by bruising your chosen herbs to release all their stored aromas. Bring 500ml (17fl oz) of cider vinegar to a gentle simmer in a pan and add the herbs. Turn off the heat and leave to cool, then strain the vinegar into a sterilized bottle (see page 53).

FRUIT VINEGARS

Many fruits can be used to flavour vinegar. Use red wine vinegar for red fruits and white wine vinegar for lighter fruits. Bring 400ml (14fl oz) of vinegar to a simmer in a pan, then add 250g (8oz) fruit and 2–3 tablespoons of sugar to the pan, bring back to the boil, then turn off the heat and leave to cool. Scoop any scum off the top and strain the vinegar through a jelly bag or fine sieve into a sterilized bottle (see page 53).

FLORAL VINEGARS

You can also use flowers, such as lavender or nasturtiums, to make unusual vinegars. Use 500ml (17fl oz) white wine vinegar for 10–15 flower heads and prepare as for herb vinegars.

RASPBERRY VINEGAR

Delicious served with orange and red onion salad or as a dressing for a bulgur wheat and pomegranate salad, this simple vinegar is incredibly versatile. Try reducing it down to intensify the flavours.

400ml (14fl oz) red wine vinegar

4 tablespoons sugar

250g (8oz) raspberries

To make the raspberry vinegar, follow the instructions opposite.

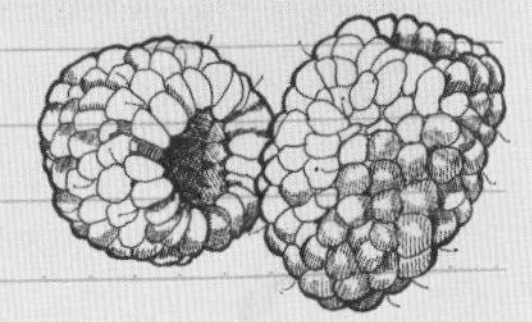

ROSEHIP VINEGAR

This flavoured vinegar goes very well with chicken or fish, and can be drizzled over a melon and prosciutto salad for some added zing.

1 litre (1¾ pints) cider vinegar

4 tablespoons sugar

50g (2oz) rosehips

peel from 1 orange

Put the vinegar and sugar into a pan and heat until dissolved. Allow to cool, then set aside. Meanwhile, thread the rosehips and pieces of orange peel alternately on to a wooden skewer. Place the skewer in a sterilized bottle (see page 53) and pour in the vinegar. Seal and leave for at least 2–3 weeks before using.

For the ultimate in flavoured vinegars, try this balsamic glaze – a reduced balsamic vinegar that has been infused with the flavour of fresh figs in the heat of a pan.

SERVES 4

- 200g (7oz) fresh figs
- 300ml (½ pint) balsamic vinegar
- a pinch of ground cinnamon
- a pinch of grated nutmeg
- 3 juniper berries
- 1 tablespoon honey (optional)

FIG BALSAMIC GLAZE

Cut the figs into halves or quarters depending on their size, then place them in a pan with all the other ingredients and heat until the vinegar starts to bubble. Reduce the heat and simmer for 15–20 minutes, or until the vinegar has reduced to a third of its original volume. Finally, sieve the mixture through a funnel and into a sterilized glass or plastic bottle (see page 53). Store in the fridge.

Place the bottle of glaze in a bowl of warm water for 5 minutes before you want to use it. This will make it more viscous and easier to pour.

A perfect partnership

The oil is infused with flavour

METHOD #10

PRESERVING IN OIL

Storing under a layer of oil is probably one of the oldest preserving techniques. It has the ability not only to keep garden produce fresh but also to create a delicious extra - infused oils. Each jar will therefore provide you with two delicious opportunities: first you can consume the preserved contents as antipasto or use them as an ingredient in another dish, then the remaining infused oil can be used in salad dressings or for cooking.

PREPARING YOUR PRODUCE

If you are planning to preserve herbs or vegetables in oil, you will want the flavour to penetrate the golden liquid. Try to bruise herbs, toast nuts and crush seeds to release their flavour prior to storing. As much as possible, try to store dry produce rather than wet. Dry the herbs or vegetables with kitchen paper or in the oven before you place them in the container. The presence of water will reduce the keeping quality of the food.

PREPARING THE OIL

Different oils have very different flavours and qualities. Olive oil works fine, but try experimenting with others: rapeseed and poppy-seed for light dishes, walnut and hazelnut for crunchy flavour, groundnut and sesame for Oriental cooking, and avocado for something a bit different.

Warming the basic oil first often helps it to absorb the flavour of the added ingredient more evenly, but you will be surprised at how easily flavour will leak into the oil molecules even if you use it cold.

PRESERVING

Put your prepared herbs or vegetables into sterilized jars or bottles (see page 53) and cover with oil. It's a good idea to use a funnel if you are filling a bottle. Ensure that you pack the container fully and, after the oil has been poured in, tap the bottom of the jar or bottle on a tea towel to dislodge any air bubbles. Seal and label.

Time is the key with oil preserving – always wait at least 3 days before using your oils, and longer for more intensity. Give the bottles a little shake during the first couple of days.

Keep the oil bottles and containers in a dark place to avoid discoloration. If you want to have your gorgeous oily collection on display, make sure it isn't in direct sunlight.

ROASTED PEPPERS IN OIL

First, cut your peppers into large slices. Put them on a baking tray and drizzle with oil. Roast them in the oven at 180°C (350°F), Gas Mark 4 for 45 minutes. Once the roasted peppers have cooled enough to handle, follow the numbered steps on this page. For added flavour when roasting the peppers, include some garlic cloves and put them into the jar with some bruised oily herbs, such as rosemary, thyme or oregano.

You can use exactly the same method to preserve aubergines, courgettes and globe artichokes.

Remove the skins, stalks and seeds from the roasted peppers. Tear or cut into slices and let dry on kitchen paper. Sprinkle with herbs.

Place the prepared vegetables and any extra flavourings, such as herbs, in a sterilized jar (see page 53).

Fill the jar with olive oil, making sure that the vegetables are completely covered. Seal and store in the fridge or a cool, dark cupboard.

HERB OILS

Scented herb oils are a shortcut to masses of flavour and will preserve the aromatic qualities of any fresh herb for longer than dried herbs in jars. These are some of our favourite herb oils, but any fresh herbs will do.

MINT OIL

300ml (½ pint) olive oil

1 teaspoon sugar

2 tablespoons chopped fresh mint

Put all the ingredients into a pan and heat for 10 minutes on a medium heat. Leave to cool, then refrigerate for 1 hour. Mint tends to discolour over time, so strain through a fine sieve and store in a sterilized glass bottle (see page 53).

ROSEMARY & GARLIC OIL

4 garlic cloves, roasted

300ml (½ pint) olive oil

3 large sprigs of fresh rosemary

Roast the garlic in a preheated oven at 180°C (350°F), Gas Mark 4 for 30 minutes. Ideally, do this when the oven is being used for something else. When the roasted garlic is cool, put it into a sterilized glass bottle (see page 53) with the oil and rosemary sprigs and leave to infuse for 2–3 days before using.

MIXED HERB OIL

2 sprigs of fresh thyme

2 sprigs of fresh oregano

2 sprigs of fresh rosemary

6 fresh basil leaves

5 pink peppercorns

300ml (½ pint) olive oil

Bruise the herb sprigs and basil leaves a little to release their flavour, and put them into a sterilized glass bottle (see page 53) with the peppercorns and olive oil. Leave to infuse for 2–3 days before using.

Flavoured oil adds something special to goats' cheese, but you can use the same method for any cheese with a similar texture and skin. This recipe turns the goats' cheese into a meal in its own right. Just spread the cheese, oil, seeds and peppercorns on crispbread or toasted granary bread and enjoy.

SERVES 4

150g (5oz) soft or semi-soft goats' cheese

1 tablespoon pink peppercorns

2 teaspoons fennel seeds

350ml (12fl oz) good extra virgin olive oil

75g (3oz) olives, dried if they have been in brine (optional)

GOATS' CHEESE WITH PINK PEPPERCORNS

Cut the goats' cheese into large pieces – if you have small rounds with a skin on, cut them so that the inside of the cheese is exposed.

Heat the peppercorns and fennel seeds in a small, dry frying pan for a minute or two, removing them from the heat when they start giving off their aroma but before they are coloured. Pour the oil into a bowl, add the warm fennel seeds and peppercorns and stir well.

Put the cheese carefully into a sterilized 500ml (17fl oz) jar (see page 53), trying not to crumble it, and adding the olives at the same time if you are using them. Tip the jar to one side and pour in the oil and spices. Make sure the oil covers the cheese – add more than listed if necessary. Put on the lid, then lay a folded tea towel on a work surface and tap the bottom of the jar on the towel to dislodge any air bubbles.

Store in a dark, cool cupboard for at least 24 hours before eating. The cheese will keep for up to 2 weeks unopened. Once opened, store in the fridge for up 10 days. Remove from the fridge an hour before you need it and serve at room temperature.

A small frying pan is ideal for dry-fying spices

All packed and ready to eat

METHOD #11

BOTTLING

Have a go at bottling if you want rows of colourful jars lined up in your cupboard and a rich supply of fruit and vegetables ready to eat over the colder months. Bottling is a relatively simple preserving method and is worth attempting at home with your own home-grown goodies. Food is placed in jars with either a syrup for fruit or a brine for vegetables, and both the jars and the food are heated until all bacteria, mould, fungi and viruses are killed. The jars are then immediately sealed. It is very important that you invest in some good-quality jars as a good seal is required to prevent spoiling.

HOW TO BOTTLE TOMATOES

Make a hot brine solution (see opposite). Pack your jars as tightly as possible with skinned tomatoes, then fill them with brine and place the lids on loosely.

Sterilize the filled jars (see opposite). After cooling, test for a good seal and vacuum by loosing the jar's clip or band and lifting the jar by the lid. If the seal does not hold, eat the contents quickly, or freeze it or try again.

PREPARING CONTAINERS

You need some bottling jars, such as Kilner or Mason jars – these are stronger than normal jam jars and have either screw tops or clips with rubber seals. Wash your jars thoroughly, then sterilize them either in boiling water for 5 minutes or in the oven (see page 53).

PREPARING SYRUP & BRINE

For fruit, make a syrup with 400g (14oz) of sugar and 1 litre (1¾ pints) of water and heat to over 60°C/140°F. Prepare the fruit by taking off stalks, rinsing clean and coring or removing stones if you wish. For vegetables, make a brine with 25g (1oz) of salt and 1 litre (1¾ pints) of water. Prepare the vegetable as required.

STERILIZING BOTTLED FOOD

Pack the fruit into the jars and top up with the hot syrup. Carefully tap the jars on a worktop to force any air bubbles to the surface and help the fruit to settle. Having removed any air gaps that could damage the fruit later, wet the rubber seals of the jars in boiling water and put them on to the lids of the jars. Seal the lids loosely and stand the jars on a wooden rack or a folded tea towel in a large pan. (If you are doing a lot of bottling it's worth investing in a specialist sterilizing pan with a false bottom for standing the jars on.) Pour in just enough cold water to reach the top of the jars, below the lids, place them on the heat and bring to a simmer according to the chart below.

SAFETY!

Acidity in food is important to prevent the *botulinum* bacteria ruining your stored produce, and to prevent serious, in some cases fatal, food poisoning. For safety's sake, you need to heat the jars beyond the boiling point of water. With low-acid foods such as vegetables, this involves using a specialist canner, which is essentially a large pressure cooker designed for bottling.

BOTTLING GUIDELINES

SOFT FRUIT

Put the jars in cold water and raise the temperature slowly over the course of an hour to 60°C (140°F). Simmer at 80°C (175°F) for a further 10 minutes.

STONE FRUIT

Repeat as above for the first hour, but then heat the water up to 85°C (185°F) and keep at this temperature for 15 minutes..

TOMATOES & GHERKINS

Repeat as above for the first hour, but then heat the water up to 90°C (195°F) and keep at this temperature for 30–40 minutes.

At the end of the summer there is a period when the new-season lambs have fattened and the chillies and figs have ripened. This dish brings together all these flavours and the mix of earthy, sweet and spice seems to us to perfectly sum up the end of summer.

SERVES 4

1 red onion, cut into chunks
1 head of fennel, thinly sliced
½ a celeriac, diced
a sprig of fresh rosemary
olive oil
800g (1lb 10oz) lamb cutlets
salt and freshly ground black pepper
25g (1oz) butter
mashed potato, to serve

FOR THE FIGS IN HONEY

6 large figs
150 ml (5 fl oz) runny honey

FOR THE SAUCE

150ml (5fl oz) port
1 fresh red chilli, seeded and chopped

HONEYED FIGS WITH LAMB

First, make the figs in honey. Simply place the fruit in a jar and pour over the honey. They will keep for 3–4 weeks.

Preheat the oven to 180°C (350°F), Gas Mark 4. Place the onion, fennel and celeriac in a roasting tray with the rosemary and a splash of olive oil. Put in the oven for 30 minutes, then add 100g (3½ oz) figs to the roasting tray and put back into the oven for a further 10 minutes.

To make the sauce, slice 100g (3½ oz) figs into small chunks and put them into a pan with the port and chilli. Bring to the boil, then reduce the heat and simmer until the sauce starts to thicken. Turn off the heat, sieve to remove the pieces of fig and chilli and set the sauce to one side.

Season the lamb cutlets all over with salt and pepper. Melt the butter in a frying pan and fry the cutlets for 2–3 minutes on each side. Remove from the pan and keep warm.

Add a little honey from the jar of figs to the juices in the pan. Stir well, then whisk into the port sauce.

Arrange the cutlets, vegetables and figs on a serving platter, pour the sauce over then serve with mashed potato.

Apricots preserved in almond-flavoured Amaretto liqueur are delicious straight from the jar. We love them with rice pudding, but this recipe goes one step further to create a very special tart.

SERVES 4

400ml (14fl oz) milk
1 vanilla pod
100g (3½ oz) round soft-grain rice
1 tablespoon crème fraîche, plus extra for serving

FOR THE AMARETTO APRICOTS

10 apricots
250ml (8fl oz) Amaretto liqueur

FOR THE PASTRY

125g (4oz) softened butter
125g (4oz) sugar
250g (8oz) plain flour
50g (2oz) ground almonds

FOR THE TOPPING

50g (2oz) butter, melted
2 tablespoons sugar

AMARETTO APRICOT TART

First make the Amaretto apricots. Halve and stone the apricots, then pack them gently into a sterilized jar (see page 53). Split the stones, put the kernels into the jar and top up with Amaretto. Store in a dark cupboard for at least a fortnight before using. They will keep for up to a year.

To make the pastry, mix the butter and sugar in a bowl with a wooden spoon. Slowly add the flour and almonds, using your hands to bring the dough together into a ball. Wrap the dough in clingfilm and store in the fridge while you make the filling.

Bring the milk to the boil, then add the vanilla pod and rice. Simmer on a very low heat for 25 minutes.

While the rice is cooking, pureé about 175g (6oz) of the apricots to a smooth paste with a tablespoon of the liqueur. You need to reserve a few apricots for decoration. When the rice is cooked, remove the vanilla pod, then add the crème fraîche and the apricot paste and stir thoroughly.

Preheat the oven to 200°C (400°F), Gas Mark 6. Take the pastry out of the fridge and roll it out to fit a 25cm (10 inch) flan tin. Prick the base with a fork, then line with greaseproof paper and baking beans. Bake in the oven for 18–20 minutes, until a pale golden colour. Remove the paper and beans and allow to cool.

Pour in the rice filling. Drizzle with the butter and sugar, then top with the remaining apricots. Cook in the oven for 20 minutes.

Serve warm, with a spoonful of crème fraîche.

Beautiful – and potent

METHOD #12

RUMTOPF

Having lived in Germany, we thoroughly enjoy the traditional German Christmas treat of ice cream with rumtopf – fruit preserved in rum. Rumtopf ('rum pot') allows you to eat out-of-season fruit all through the year, and is particularly popular over the festive season.The fruit and liquor from a rumtopf are great with cream or ice cream, and you can add the liquor to sparkling wine to make a cocktail, or drink a small glass of it on its own. The content of a rumtopf is incredibly alcoholic – you have been warned!

BEGINNING YOUR RUMTOPF

A rumtopf made in an authentic crock will give you up to 50 servings. If you can't find an authentic crock, you can use a small lidded crock with a volume of 2–3 litres.

Clean the crock, dry it and start it off with 400g (13oz) of any fruit except apples and 200g (7oz) of sugar. Cover the fruit and sugar with rum, put clingfilm over the top of the rumtopf and put the lid on.

ADDING TO YOUR RUMTOPF

Every time a new fruit comes into season, open the crock, put the fruit in, add more sugar and cover with rum again. We add 400g (13oz) of each fruit at a time, plus half as much sugar, and continue adding fruit until the rumtopf is full. We usually start with strawberries and then, as the seasons develop, we add cherries, plums, blackberries and raspberries. Once the last fruit has been added, leave undisturbed until Christmas.

FESTIVE RUMTOPF

For each addition to the crockpot:

400g (13oz) of each type of soft fruit, whatever is in season (we use smallish strawberries, plums, cherries, blackberries and raspberries)

200g (7oz) sugar for each 400g (13oz) fruit

strong brown rum, enough to cover the fruit (extra rum will be needed later in the year)

To preserve your fruit in the rumtopf, follow the instructions opposite.

HOW TO MAKE A RUMTOPF

Add the fruit to the crockpot as each season arrives. Hull the strawberries, stone and halve the plums, pit the cherries, and so on.

Put the fruit and sugar into the crockpot and cover with rum.

You need to keep the fruit underneath the liquor, so pop a plastic lid or saucer into the crock to hold it down.

Put clingfilm over the top of the crockpot and put the lid on. Every time a new fruit comes into season, open the crock, put the fruit in, add more sugar and rum and cover again.

METHOD #13

RELISHES

There is a thin line between a chutney and a relish. The cooking process is almost identical, and the textures are similar, but in our opinion there is a substantial difference in taste. If you make a relish, there is more scope for it to be sweeter, fruitier or have a potent mustard quality. Relishes are usually highly flavoured, often spicy hot or sweet and sour. We find that a relish comes out of the cupboard with cheeses and whenever we eat homemade burgers.

PREPARING INGREDIENTS

As with many preserves, the first stage involves chopping your fruit and vegetables so that they are ready for the process. With a relish, the general rule is to aim for small cubes that will spread nicely once cooked. If you put the chopped ingredients into a large glass bowl, you can see the quantities of the various ingredients. This will give you a good idea of the eventual taste and allows you to adjust the proportions as you like.

COOKING THE RELISH

Start by dissolving the sugar in the vinegar in a heavy-bottomed pan on a medium heat. When it starts to simmer, add the fruit and vegetables and any spices. Bring to the boil, then return it to a simmer and cook for 30–45 minutes. When some of the vinegar starts to reduce and you are happy with the consistency, take the pan off the heat and pour the relish straight into sterilized jars (see page 53) – a funnel is useful here. A relish can be pulpy and set, more like a jam, or it can be juicy. If you want it to end up firm, make sure you cook it for longer.

FINISHING IT OFF

Near the end of the cooking process you can add some small ingredients, such as chopped gherkins or pickled onions. This will improve the texture of your relish and give it a crunch and zingy flavour.

STORING YOUR RELISHES

Seal the jars straight away and store them in a cool, dark place for up to a year. Once opened, keep refrigerated.

TOMATO RELISH

Makes 2kg (4lb)

250g (8oz) brown sugar
500ml (17fl oz) vinegar
1kg (2lb) red tomatoes, finely chopped
150g (5oz) onions, chopped
3 red chillies, chopped
2 tablespoons mustard seeds
1 teaspoon paprika
1 teaspoon ground ginger
salt and freshly ground black pepper
100g (3½oz) gherkins, finely chopped
100g (3½oz) pickled onions, finely chopped

Put the sugar and vinegar into a large heavy-bottomed pan and heat until dissolved. Bring to a simmer and stir in the tomatoes, onions, chillies and spices. Cook on a medium heat for 30 minutes, then add the gherkins and pickled onions. Simmer for a further 5 minutes, then pour into sterilized jars (see page 53).

JALAPEÑO RELISH

Makes 1kg (2lb)

125g (4oz) light brown sugar
250ml (8fl oz) cider vinegar
500g (1lb) fresh jalapeño chillies, seeded and cut into 5mm (¼ inch) strips
150g (5oz) onions, cut into 5mm (¼ inch) strips
150g (5oz) carrots, cut into 5mm (¼ inch) strips
1 teaspoon fennel seeds
1 teaspoon brown mustard seeds
1 teaspoon yellow mustard seeds

Put the sugar and vinegar into a large heavy-bottomed pan and heat gently until the sugar has dissolved. Put the chillies, carrot and onion into the vinegar solution and bring to the boil. Add the spices and simmer for 30 minutes, then pour into sterilized jars (see page 53).

This sweetcorn relish is incredibly versatile, but we think it's especially good with pork, beans and a green salad. The further ahead you make the relish, the better, as storage improves the intensity of the flavours.

SERVES 4

400g (13oz) shoulder of pork, diced

4 tablespoons honey

1 fresh red chilli, seeded and chopped

1 teaspoon paprika

a pinch of cayenne pepper

2 tablespoons orange juice

1 teaspoon fennel seeds

FOR THE SWEETCORN AND PEPPER RELISH

1 red onion, finely chopped

1 garlic clove, chopped

1-2 teaspoons sunflower oil

2 large cobs of sweetcorn

1 red pepper, finely chopped

1 fresh red chilli, seeded and finely chopped

juice of 1 lime

2 shots of tequila

50ml (2fl oz) white wine vinegar

100g (3½ oz) sugar

1 tablespoon chopped fresh coriander leaves

TO SERVE

flatbread

refried beans

fresh salad leaves

SPICED PORK WITH SWEETCORN & PEPPER RELISH

First make the relish. Put the onion and garlic into a hot pan with the sunflower oil and soften without colouring on a low heat. Slice the corn kernels off the cob with a sharp knife and add to the pan with the pepper, chilli and lime juice. Pour in the tequila and simmer for 2–3 minutes to burn off the alcohol, then add the vinegar and sugar. Heat for 10–15 minutes, stirring to prevent the mixture sticking to the base of the pan. Add the coriander leaves and pour into a sterilized jar (see page 53) Set aside until you are ready to serve. The relish will keep in the fridge for 4–6 weeks, but once opened, use within a week.

Put the pork into a bowl and add the rest of the ingredients. Leave to marinate for at least an hour. Meanwhile, soak 8 wooden skewers in cold water.

Take the pork out of the marinade and cook it for 2–3 minutes on each side, either on skewers over a flame or in a griddle pan, brushing with the marinade during the cooking for extra stickiness.

Serve with flatbread, refried beans and a fresh green salad.

Cucumber and dill are a bit like tomato and basil – their compatibility is unquestionable. Together they make a cooling relish that's perfect with a piece of fish.

SERVES 4

100g (3½ oz) breadcrumbs
1 tablespoon finely chopped fresh dill
zest of 1 lemon
1 teaspoon sea salt
freshly ground black pepper
2 tablespoons olive oil
4 x 200g salmon fillets

FOR THE RELISH

1 cucumber
2 tablespoons chopped fresh dill
1 tablespoon white wine vinegar
2 teaspoons sugar
1 teaspoon white mustard seeds (optional)
sea salt

FOR THE VEGETABLES

100g (3½ oz) cherry tomatoes
2 red peppers, cut into strips
olive oil
250g (8oz) new potatoes, halved
20g (¾ oz) butter
4 spring onions, finely chopped
250g (8oz) green beans

BAKED SALMON WITH CUCUMBER & DILL RELISH

Start by making the relish. Grate the cucumber into a pan and stir in the chopped dill. Add the vinegar, sugar and mustard seeds, if using, and stir over a low heat for 5 minutes. Season with a pinch of salt and set aside to cool. Stored in a sterilized jar (see page 53), the relish will keep in the fridge for up to 3 weeks.

Preheat the oven to 200°C (400°F), Gas Mark 6. Put the breadcrumbs into a bowl and add the dill, lemon zest, salt and pepper. Stir in the oil, then spread the mixture over the fillets and lay them on a baking tray.

Put the tomatoes and peppers alongside the fillets if there is room, otherwise use a separate roasting tin. Drizzle with olive oil and sprinkle with salt and pepper, then put the tray(s) into the oven and cook for 20 minutes.

Meanwhile bring a large pan of water to the boil. Cook the potatoes for 15–20 minutes. When they're ready, drain and return them to the pan with a knob of butter and the spring onions. Smash them up a bit with a fork. Blanch the green beans in boiling water for 3–4 minutes.

Serve the salmon fillets with the roasted vegetables, smashed potatoes, green beans and cucumber relish.

Duck and orange are renowned companions, since the acidity of the fruit cuts through the natural richness of the meat. This recipe adds vinegar and spices to the mix and makes a great accompaniment to cold duck or goose.

MAKES 600g (1¼lb)

- 5 large oranges
- 2 tablespoons extra virgin olive oil
- 1 large onion, chopped
- 2 Bramley apples, peeled and diced
- seeds from 8 cardamom pods
- 1 teaspoon black mustard seeds
- ½ teaspoon ground cloves
- 150g (5oz) light brown sugar
- 300ml (½ pint) cider vinegar

SPICY ORANGE RELISH

Use a vegetable peeler to take off the zest of the oranges in pieces. Cut the zest into strips about 5mm (¼ inch) wide and set aside. With a sharp knife, cut the pith and skin off the oranges and take out the flesh in segments.

Heat the oil in a pan, then add the onion and cook until softened. Add the apples, cardamom and mustard seeds and cook for a further 5 minutes.

Add the orange segments and zest to the pan with the remaining ingredients and simmer for 40 minutes. Put into sterilized jars (see page 53) and allow to cool. Store the relish in the fridge, where it will keep for up to 6 months.

Serve with cold duck or goose, as a salad or in a sandwich.

you need segments without pith or peel

you have to try this

METHOD #14

CHUTNEY

Chutney can be made from almost any fruit or vegetable. We've used marrows, runner beans, apples, radishes, rhubarb, red and green tomatoes, grapes, chillies, aubergines, squash, pears and turnips, to name but a few! The key when making chutney is to cook it for a long time and evaporate most of the moisture so that it reaches a thick, jam-like consistency. The colours will change but the flavours will always intensify. The other thing to consider when cooking a good chutney is to be bold and contrast the flavours - don't be afraid to mix fruits and vegetables.

CHOOSING FLAVOURINGS

There are various spices that add those distinctive chutney flavours: cumin, coriander, allspice, cloves, ginger, peppercorns, paprika, mustard seed and garlic are some of the main ones. Crush your chosen spices with a pestle and mortar to release their aroma.

Vinegar is used in chutney because its acidity effectively inhibits the actions of unwanted micro-organisms. Distilled vinegar is the strongest preservative, but also one of the most expensive. The more you pay for your wine or malt vinegar, the tastier your chutneys will be. We tend to use a variety of different vinegars, including cider vinegar, white wine vinegar and red wine vinegar.

STORING

Most chutneys mature with age, and if kept in the right conditions will last for years, but once you have opened a jar of chutney, it will last longer if you keep it in the fridge.

GRANDPA'S RECIPE

Makes about 12 small jars

- 2.5kg (5lb) green tomatoes, sliced
- 500g (1lb) onions, finely chopped
- 1 tablespoon salt
- 500g (1lb) cooking apples, peeled, cored and sliced
- 500g (1lb) sultanas, chopped
- 1 litre (1¾ pints) pickling vinegar
- 500g (1lb) light muscovado sugar
- 5 small hot chillies, finely chopped
- 2 tablespoons ground ginger

Put the tomatoes and onions in a bowl, sprinkle with the salt and set aside for a few hours. Drain off the liquid and place the solids in a pan with the apple and sultanas. Gently heat until the fruit softens. Add the vinegar, sugar, chillies and ginger and cook for at least 45minutes. Pour into sterilized jars (see page 53) and seal.

HOW TO MAKE CHUTNEY

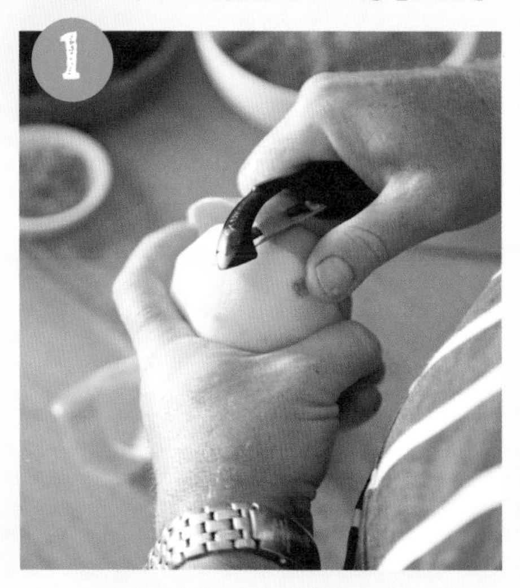

Start by slicing, dicing, peeling and coring all the fruit and vegetables you will be using. Then prepare the spice or herb mix.

Add the fruit, vegetables, herbs and spices to a pan and simmer until they start to soften.

Cook on a medium to low heat for no less than 45 minutes. You will know when your chutney is ready because it will have reduced to a jam-like consistency and your kitchen will smell wonderful.

Remove the chutney from the heat and pour into sterilized jars (see page 53). Keep the jars in a cool, dark place and make sure that all the lids are securely sealed.

Chutney and cheese is a partnership that needs no introduction. Spicy chutneys are not to everyone's taste, but for those of us who enjoy a bit of heat, this preserve is hard to beat.

SERVES 4

- 1 small baguette
- 1 garlic clove, crushed
- 2-3 tablespoons olive oil
- 300g (10oz) goats' cheese
- 200g (7oz) mixed bitter salad leaves
- 1 carrot, grated

FOR THE DRESSING

- 1 teaspoon lemon juice
- 1 tablespoon olive oil
- 1 teaspoon finely chopped fresh ginger
- 1 fresh red chilli, seeded and finely chopped
- 1 teaspoon sesame seeds

FOR THE CHUTNEY

- 3 fresh red chillies, chopped
- 2 large cooking apples, peeled, cored and chopped
- 450g red tomatoes, chopped
- 2 red onions, finely chopped
- 200g (7oz) sugar
- 250ml (8fl oz) cider vinegar
- 100g (3½oz) raisins, chopped
- a pinch of cayenne pepper
- a pinch of ground allspice
- 1 teaspoon chopped ginger
- salt and freshly ground black pepper

GOATS' CHEESE CROSTINI WITH 'SOME LIKE IT HOT' CHUTNEY

To make the chutney, put all the ingredients into a large pan, bring to the boil, then reduce the heat and simmer for 45 minutes. When the smells have filled your kitchen, the fruit has all softened and the vinegar has reduced, transfer the chutney into sterilized jars (see page 53) and store in a dark, cool place, where it will keep for 6–12 months. Allow the chutney to mature for a month before eating.

Slice the bread, then toast it until golden. Rub a little crushed garlic and olive oil over one side of each piece of toast and spread some goats' cheese on top.

Combine the bitter salad leaves and grated carro in a bowl. To make the dressing, put the lemon juice, oil, ginger, chilli and sesame seeds into a bowl and mix with a fork. Drizzle over the salad and serve with the crostini, topped with a dollop of chutney.

you can never have too much jelly

METHOD #15

JAMS & JELLIES

Jams and jellies are made in a similar way: both involve heating fruit with sugar to activate the pectin (a natural setting agent) found in a variety of ripe fruits and vegetables, then storing it in a suitable container. Some fruit doesn't contain very much pectin, so you might need to add extra pectin, which you can buy in liquid form, to some jams and jellies. What's the difference between a jam and ajelly? If there are any fruity bits in it, it's a jam. If not, it's a jelly.

GUIDELINES

- Always use undamaged fruit. Fruit that is damaged can spoil the flavour and the jam is likely to deteriorate quickly.
- If fruits are low in pectin, then either fruits with a higher level of pectin need to be included, or liquid pectin should be added.
- Always make sure the sugar is completely dissolved before bringing to the boil or the result can be grainy.
- Ensure all equipment you use is clean (for jelly-making, always boil-wash the muslin, jelly bag or tea towel before using).
- Use a non-reactive pan (preferably stainless steel), and wooden or plastic spoons.
- Don't make too large a quantity at once or it can take a long time to come to the boil and the fruit can break up.
- Skim the scum that forms once the setting point has been reached.
- Allow the jam or jelly to stand for 15 minutes after a set has been reached before putting it into jars. This will aid the distribution of the fruit within the jars.

PECTIN GUIDELINES

	PECTIN LEVEL
CITRUS PEEL	30%
ORANGES	0.5–3.5%
APPLES	1–1.5%
CARROTS	1.4%
APRICOTS	1%
CHERRIES	0.4%

- Store unopened jams and jellies in a cool, preferably dark place. Only store in the refrigerator once opened.

CHOOSING FRUITS

Not all fruits contain sufficient pectin to make jam or jelly successfully. Soft fruits, such as cherries, grapes and strawberries, contain small amounts of pectin, so to use them successfully you will need to add extra pectin to your recipe or use preserving sugar, which already has pectin in it. Apples, quinces, plums, gooseberries, blackberries, redcurrants, oranges and other citrus fruits contain large amounts of pectin.

MAKING JAM

Jams tend to be made with fruit that has been cut into pieces or crushed, then heated with water and sugar. A good jam should have an even consistency – it should spread easily and have no free liquid.

MAKING JELLY

Jellies are clear or translucent fruit spreads and can be sweet, savoury or spicy. The process for making a jelly is similar to that for making jam, but there is an additional step where the fruit pulp is strained out using a jelly bag.

A good jelly should quiver when moved but hold an angle when cut.

Testing for setting

TESTING FOR SETTING POINT

Place a small plate in the fridge or freezer for 15 minutes. Pour a spoonful of the hot mixture onto the chilled plate and return it to the fridge for 5 minutes. To test, push the edge of the mixture with your index finger – it is ready when it wrinkles. If not, continue to cook, checking for setting point every few minutes. It is tempting to keep cooking to achieve a firmer set, but overcooking can be detrimental to the flavour.

STORING

All jams and jellies must be put into sterilized jars (see page 53) and sealed prior to storage. Jars with click-down lids (where the top of the lid bends inwards as the air above the jam cools and contracts) are a good option as it is easy to tell if the seal has been maintained during storage. If the jam or jelly looks at all fizzy or mouldy, the seal has failed.

HOW TO MAKE STRAWBERRY JAM

Makes 3.25kg (7lb)

2kg (4lb) strawberries

2kg (4lb) preserving sugar

juice of 3 lemons

To make the strawberry jam, follow the numbered steps on this page and opposite.

Hull the strawberries and place them in a large non-reactive bowl (stainless steel, china or glass). Sprinkle over the sugar and lemon juice. Leave in the fridge overnight.

Turn off the heat and test for setting point (see page 93). If the jam is not setting, bring it back to the boil for 2 minutes and retest.

Once setting point has been reached, allow to cool for 5 minutes, then skim off any scum.

Put a small plate in the fridge to chill. Tip the contents of the bowl into a large, heavy-bottomed pan and heat gently until all the sugar has dissolved. Turn up the heat and boil vigorously for 5 minutes.

After a further 10 minutes, put the jam into sterilized jars (see page 53).

GOOSEBERRY JAM

Makes 5.5kg (12lb)

2.5kg (5lb) gooseberries
1 litre (1¾ pints) water
3kg (6¾ lb) preserving sugar

Put the gooseberries and water in a large, heavy-bottomed pan and gently heat until the fruit is soft. Add the sugar and heat until dissolved. Turn up the heat and boil rapidly for about 20 minutes, until the jam passes the test for setting point (see page 93). Skim off any skum and pour the jam into sterilized jars (see page 53).

APRICOT JAM

Makes 5kg (11lb)

2.5kg (5lb) apricots
500ml (17fl oz) water
3kg (6¾ lb) preserving sugar

Cut the apricots in half and remove the stones. Place the stones on a square of muslin or in a jelly bag and tie into a bundle.

Put the apricots and water in a large, heavy-bottomed pan and gently heat until the fruit is tender. Add the sugar and heat until dissolved. Add the bag of apricot stones to the pan, turn up the heat and boil rapidly for about 25 minutes, until the jam passes the test for setting point (see page 93). When ready, pour the jam into sterilized jars (see page 53).

HOW TO MAKE ROSE PETAL JELLY

Makes 3.75kg (8lb)

2kg (4lb) Bramley apples

approx. 3kg (6lb) sugar

4 drops of edible rose essence

1 tablespoon dried edible rose petals

To make rose petal jelly, follow the numbered steps on this page and opposite.

Put a small plate in the fridge to chill. Cut up the apples, place them in a large pan and add just enough water to cover. Bring to the boil and simmer for 20 minutes.

Add 1kg (2lb) of sugar per 500ml (17fl oz) of apple liquor. Slowly bring to the boil, stirring to ensure the sugar dissolves. Boil for 5 minutes then cool for 3–4 minutes.

Skim off any scum. Boil again for 5 minutes, leave to cool for 3–4 minutes, then skim again. Test for setting point (see page 93). If not reached, boil for a further 5 minutes.

Strain the liquid through muslin or a jelly bag. Don't squeeze the bag or you will end up with cloudy jelly. Measure the liquid and put it into a large, heavy-bottomed pan.

Add the rose essence and petals and stir in. When the jelly has cooled slightly, pour into sterilized jars (see page 53), adding a bit at a time to each jar. As the jars cool, rotate them to ensure that the petals do not sink or float.

REDCURRANT JELLY

Makes 2.75kg (6lb)

3kg (6lb) redcurrants

500g (1lb) preserving sugar per 500ml (17fl oz) juice)

Put a small plate in the fridge to chill. Place the redcurrants in an unsealed glass jar. Place the jar in a pan of cold water and bring to a simmer over a low heat. Simmer for about an hour, mashing the berries from time to time.

Strain the fruit through muslin or a jelly bag, allowing it to drain overnight. Measure the liquid and put it into a large, heavy-bottomed pan. Add 500g (1lb) of preserving sugar per 500ml (17fl oz) redcurrant juice.

Turn up the heat and boil for 5 minutes. Test for setting point (see page 93). When ready, skim off any scum and pour the jam into sterilized jars (see page 53).

Cream teas are a ritual that just about everyone is familiar with. That said, the big debate is whether you put the jam or the cream on the scone first. For us it doesn't really matter, as long as there is plenty of both!

MAKES 10–12

- 450g (14½ oz) self-raising flour
- ¼ teaspoon salt
- 100g (3½ oz) unsalted butter, cut into small cubes
- 75g (3oz) caster sugar
- 250ml (8fl oz) buttermilk
- 2 teaspoons vanilla extract
- 1 egg, beaten
- 2 tablespoons milk

TO SERVE

- clotted cream
- strawberry jam

THE ULTIMATE CREAM TEA

Preheat the oven to 220°C (425°F), Gas Mark 7. Grease and flour a baking tray, or use a silicone sheet.

Place the flour, salt, butter and sugar in a large bowl and rub together with your fingertips until all the butter has been incorporated. In a smaller bowl or a jug, combine the buttermilk, vanilla and most of the egg. Gradually pour the buttermilk mixture into the large bowl, mixing gently with your hand. It's important not to overmix as that will knock the air out of it.

Turn the mixture on to a work surface and shape it into a circle about 4cm (1¾ inches) thick. Using a cutter about 6cm (2½ inches) in diameter, stamp out circles and place them on the prepared baking tray. Mix the milk with the remaining egg and use this to glaze the top of the scones.

Bake in the oven for about 15 minutes, until risen and golden brown. When ready, transfer them to a rack for a about 10 minutes. Slice in half, spread with plenty of cream and jam and enjoy with a cup of tea.

For a really fancy-looking fruit tart, apricot jam is essential. It glazes the fruit – keeping it moist and bright – and flavours the pastry while also preventing sogginess or drying.

SERVES 4

FOR THE PÂTE SUCRÉE

50g (2oz) sugar

100g (3½ oz) butter, softened

1 egg, beaten

200g (7oz) plain flour

a pinch of salt

FOR THE APRICOT GLAZE

100g (3½ oz) apricot jam

1 tablespoon water

1 tablespoon Grand Marnier liqueur

FOR THE CUSTARD

50g (2oz) sugar

3 large egg yolks

20g (¾ oz) flour

20g (¾ oz) cornflour

300ml (½ pint) milk

1 teaspoon vanilla extract

FOR THE FRUIT

100g (3½oz) blackberries

100g (3½oz) raspberries

100g (3½oz) strawberries, sliced

1 kiwi fruit, sliced

TUTTI FRUITY TART

To make the pâte sucrée, beat the sugar and softened butter together in a bowl and gradually add the egg, flour and salt. Form into a ball, wrap in clingfilm, and put into the fridge for 30 minutes.

Meanwhile, make the apricot glaze. Heat the apricot jam, water and Grand Marnier in a saucepan until smooth, then strain through a fine sieve to remove any lumps. Leave to cool while you make the custard.

Mix the sugar and egg yolks together in a bowl, then sift in the flour and cornflour. Stir until you have a smooth paste. Heat the milk in a pan with the vanilla until it just starts to boil, then remove from the heat and add it gradually to the flour mixture, stirring slowly. When all the milk is incorporated, pour it back into the pan and return it to the heat for another few minutes, whisking constantly until the custard comes to the boil and thickens. Turn off the heat, cover and set aside to cool.

Preheat the oven to 180°C (350°F), Gas Mark 4 and grease a 20cm (8 inch) cake tin. Take the pastry out of the fridge and roll it out to fit the tin. Lay the pastry in the tin, prick the base with a fork, then line with greaseproof paper and baking beans. Bake for 12–15 minutes, then remove the paper and beans. Allow to cool.

Brush inside the pastry case with some of the glaze and allow it to dry. Pour in the cooled custard. Arrange the fruit on top in overlapping concentric circles. Warm the rest of the glaze and brush over the fruit.

This is a delicious way to serve game meat. Wrapping it in sausage meat helps it stay moist, so tender game that usually suffers from drying out, such as pigeon, pheasant and rabbit, can be used in this little roast.

SERVES 4

- 8 slices of air-dried ham
- 500g (1lb) venison sausage meat
- approx. 500g (1lb) tender game meat (e.g. 2 pigeon breasts, 1 pheasant breast, 2 rabbit loin fillets)
- 8 tablespoons redcurrant jelly (see page 97)
- ½ a shallot, finely diced
- 250ml (8fl oz) port
- 50g (2oz) butter

POACHER'S SUPPER WITH REDCURRANT SAUCE

Preheat the oven to 180°C (350°F), Gas Mark 4.

Lay 2 slices of ham on a work surface so that they slightly overlap, trimming off the edges to form a square. Do the same with the rest of the ham slices. Divide the sausage meat into 4 portions, flatten them out, and lay one portion on each square of ham. Cut your game meat into strips about 5mm (¼ inch) thick and arrange them along the centre of the sausage meat at right angles to the overlap in the ham slices.

Spread about a tablespoon of redcurrant jelly over the meat and roll up each square of ham so you have game and jelly surrounded by sausage meat and wrapped in ham. Secure the rolls with cocktail sticks, place them on a baking tray and cook in the oven for 20 minutes, then rest for 5 minutes.

While the meat is cooking, make the sauce. Simmer the shallot in the port until reduced and sticky. Stir in about 4 tablespoons of redcurrant jelly and, when it is heated through, take it off the heat and stir in the butter.

Cut the meat into discs and serve with the sauce poured over. Root vegetables and mash are good accompaniments.

be generous with the jelly

A savoury jelly not only adds flavour to a dish, it will also contrast nicely with the other textures. This recipe transforms fresh rosemary into the star of the dish instead of just the background flavour.

SERVES 4

- 20g (¾ oz) butter
- 350g (11½ oz) oyster mushrooms, sliced
- 1 sprig of fresh rosemary
- 2 garlic cloves, chopped
- 1 teaspoon lemon juice
- 100ml (3½fl oz) sherry
- 100ml (3½fl oz) chicken stock
- salt and freshly ground black pepper
- 2 balls of mozzarella cheese

FOR THE ROSEMARY JELLY

- 2 large sprigs of fresh rosemary
- 250g (8oz) cooking apples, chopped
- 100ml (3½fl oz) water
- 1 garlic clove (optional)
- 100ml (3½fl oz) white wine vinegar
- 2 tablespoons sugar
- 2 tablespoons chopped fresh rosemary

MOZZARELLA & MUSHROOMS WITH ROSEMARY JELLY

Start by making the jelly. Put the rosemary sprigs, chopped apples and water into a pan and cook over a low heat for 30 minutes – you can add a clove of garlic at this point if you like. Add the vinegar, boil hard for 5 minutes, then spoon the mixture into a jelly bag suspended over a bowl and leave for a few hours or overnight.

Put a small plate in the fridge to chill. When all the liquid has dripped through the bag, put it into a pan with the sugar and chopped rosemary. Place on a high heat for 5 minutes, or until setting point is reached (see page 93). Skim off any scum and pour the mixture into sterilized jam jars (see page 53) and place them in the fridge for 5–10 minutes, removing them before the jelly fully sets. Cover and return to the fridge for a further hour, until set.

Melt the butter in a pan, toss in the mushrooms with the rosemary, garlic and lemon juice, and cook for 5 minutes. Add the sherry and let the mushrooms cook until the alcohol has evaporated. Remove the mushrooms from the pan and divide between 4 plates.

Add the chicken stock to the mushroom juices and reduce the liquid for a couple of minutes on a high heat. Season with salt and pepper, then strain the sauce and pour around the mushrooms.

Set half a mozzarella on each plate, and top with a spoonful of jelly.

A classic trifle contains sponge fingers, sherry, fruit, jelly, custard and cream. This version takes trifle to another level. It may seem that there is only a thin layer of jelly, but the flavour is there in abundance.

SERVES 8-10

- ½ a packet of sponge fingers
- 125g (4oz) Amaretti biscuits
- 150ml (5fl oz) sherry
- 450g (14½oz) fresh strawberries
- 75g (3oz) toasted almond flakes
- 500ml (17fl oz) custard

FOR THE STRAWBERRY AND CHAMPAGNE JELLY

- 1kg (2lb) strawberries, hulled and quartered
- 250ml (8fl oz) champagne
- 1kg (2lb) preserving sugar, or 1kg (2lb) granulated sugar plus 1 tablespoon fruit pectin

FOR THE TOPPING

- 450ml (¾ pint) double cream
- 50g (2oz) caster sugar
- 125ml (4fl oz) champagne

STRAWBERRY & CHAMPAGNE TRIFLE

First make the jelly. Put a small plate in the fridge to chill. Place the strawberries and champagne in a large pan and heat until the fruit is soft and the total volume has reduced by about one-third. Add the sugar and stir until it has dissolved, then simmer for about 20 minutes, until setting point is reached (see page 93). If necessary, simmer for a bit longer. When the jelly is ready, strain it through a jelly bag into sterilized jars (see page 53) and keep in the fridge for up to 3 months.

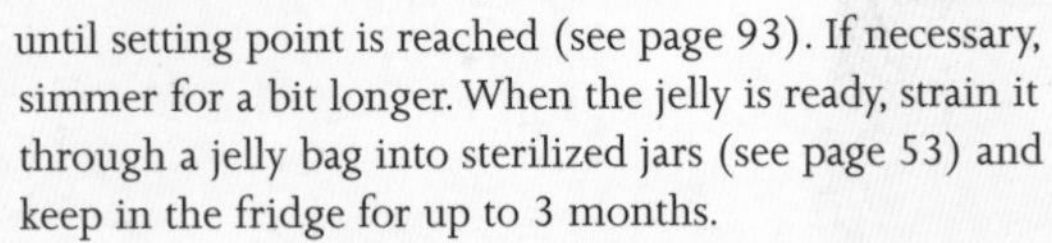

Line a glass bowl with the sponge fingers and amaretti biscuits and sprinkle them with the sherry. Warm 150–250g (5–8oz) of jelly in a small pan until it's runny, then pour it over the biscuits. Arrange the fruit and all but a small handful of almonds on top of the jelly, then pour over the cold custard.

To make the topping, whip the cream and sugar to soft peak stage, then gently fold in the champagne. Spoon over the custard and sprinkle the remaining almonds on top.

Pour sherry over the jelly

Making jelly from crab apples is a massive treat. In fact, we'd go so far as to say that crab apple jelly is probably one of the best-kept secrets in the preserving world. We use it with game as a replacement for redcurrant jelly.

SERVES 2-4

1 teaspoon chopped fresh thyme
1 teaspoon chopped fresh parsley
250g (8oz) sausage meat
zest of 1 lemon
75g (3oz) apricots, chopped
150g (5oz) chestnuts, chopped
1 cock pheasant, boned
75g (3oz) butter
4 rashers of bacon
1 tablespoon flour
150ml (5fl oz) chicken stock
salt and freshly ground black pepper

FOR THE CRAB APPLE JELLY

2kg (4lb) crab apples
500g (1lb) sugar
juice of 1 lemon

TO SERVE

peas
roasted vegetables

STUFFED PHEASANT WITH CRAB APPLE JELLY

First make the jelly. Put a small plate in the fridge to chill. Place the fruit in a saucepan and cover with water. Bring to the boil, then simmer for 30 minutes, until the fruit starts to soften. Strain through a jelly bag for a few hours or overnight. When all the liquid has dripped through the bag, put it into a pan with the sugar and lemon juice and place on a medium heat. Bring to the boil, then simmer for about 45 minutes. Check for setting point (see page 93), then skim off any scum. Pour the jelly into sterilized jars (see page 53), seal and refrigerate. It will keep for 2–3 months.

Preheat the oven to 200°C (400°F), Gas Mark 6. Put the herbs, sausage meat, lemon zest, apricots and chestnuts into a bowl and mix together. Stuff your pheasant with the mixture and put it into a roasting tray on a rack. Smear the butter over the bird and cover the breast with the bacon rashers. Roast for 45 minutes, then transfer the bird to a serving plate to keep warm.

Stir the flour into the pheasant juices in the tray, then mix in the stock and 2 tablespoons of the crab apple jelly. Place on a high heat for 2–3 minutes, stirring constantly with a whisk. Season the sauce with salt and pepper.

Serve the pheasant with peas, roasted vegetables and the sauce.

Perfect for breakfast →

METHOD #16

MARMALADES

There is nothing quite like homemade marmalade. Traditionally, Seville oranges are used, as they are bitter and, when combined with sugar, give a really intense orangey flavour. That said, the season for Seville oranges is short, so it makes sense either to make your year's supply then or to freeze the oranges and use them at your convenience. Of course, it is possible to make a really interesting marmalade all year round using any combination of shop-bought citrus fruits. Lemon and lime are definitely worth trying, but be prepared to simmer lemon and lime peel for a lot longer than orange peel to soften it – up to 90 minutes.

HOW TO MAKE MARMALADE

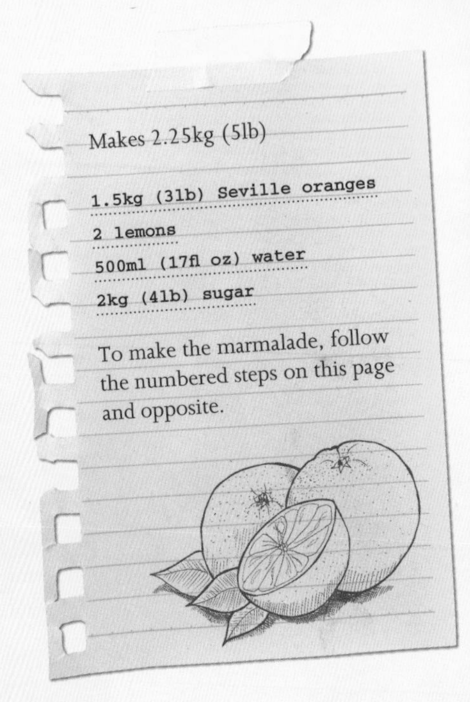

Makes 2.25kg (5lb)

1.5kg (3lb) Seville oranges
2 lemons
500ml (17fl oz) water
2kg (4lb) sugar

To make the marmalade, follow the numbered steps on this page and opposite.

First, scrub the oranges. Peel them, using a potato peeler or a sharp knife, and cut the peel into fine strips. Place a sieve over a large bowl and squeeze as much juice as possible out of the oranges and one of the lemons.

Keep all the skin, pith, pips and flesh from the sieve and place it, along with the peel and pith from both lemons, on a square of muslin or in a jelly bag, and tie into a bundle. Put a small plate in the fridge to chill.

Put the orange peel, juice and water into a large stainless steel pan and hang the bag of bits in it. Bring to the boil and simmer for 30 minutes, uncovered, until the peel is soft. Remove the bag and place it on a plate to cool.

When cool enough to handle, squeeze as much out through the muslin and into the pan as you can. Add the sugar and stir until dissolved. At this stage you can taste and add more sugar if necessary.

Put the pan of marmalade on a medium heat and allow it to boil rapidly for 15 minutes. Skim off any scum. Test for setting point (see page 93). When ready, pour into sterilized jars (see page 53), then seal and store.

Marmalade can be used for a lot more than just spreading on your toast. The bitterness of the orange rind cuts through the richness of the bread and butter pudding beautifully, but you can make this dish with any citrus marmalade. While this pudding is gorgeous on its own, you can pour some double cream over it if you fancy.

SERVES 6

- 8 slices of brioche loaf, crusts removed
- 50g (2oz) softened butter
- 8 tablespoons orange marmalade
- 3 eggs
- 4 tablespoons caster sugar
- 1 teaspoon vanilla extract
- 500ml (17fl oz) single cream
- 2 tablespoons whisky

MARMALADE BREAD & BUTTER PUDDING

Preheat the oven to 160°C (325°F), Gas Mark 3.

Butter the slices of brioche and make 4 sandwiches, each filled with at least 1 tablespoon of marmalade. Cut the sandwiches into triangles and butter the outsides, then lay them in rows in a suitable sized baking dish.

Beat the eggs in a bowl and add the sugar, vanilla, cream and whisky. Beat again, then pour over the brioche sandwiches. Allow to stand for at least 20 minutes, then use the remaining marmalade to make little orangey blobs all over the pudding.

Bake in the oven for about 45 minutes, until golden. Cool for 5 minutes before serving.

who needs sweets?

METHOD #17

CANDIED PEEL

This method creates perhaps the prettiest of our preserved treats. Candied citrus peel is colourful, sweet and perfect for decorating cakes – and the process is surprisingly easy. The important thing is to make sure the fruit you use is wax-free. Here we have candied the peel of grapefruit, oranges, limes and lemons. Other fruit, such as cherries and pineapple, will crystallize in a similar way to the peel. Follow the same method of cooking in a syrup, then another increasingly strong syrup, and finally making sure that they are dry. Experiment and have fun – you'll be well rewarded.

HOW TO MAKE CANDIED PEEL

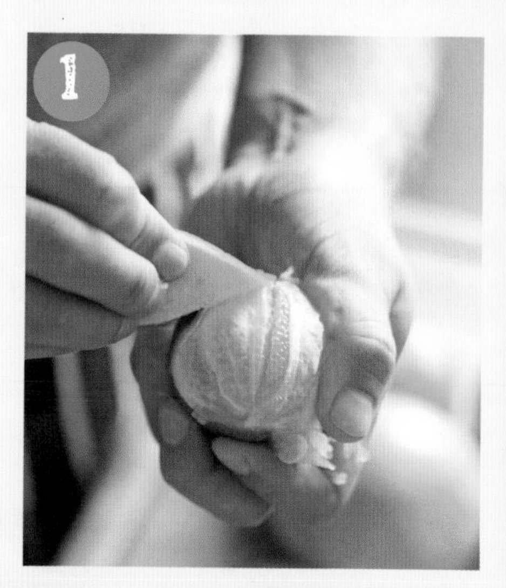

Start by preparing the peel of your chosen citrus fruit. Use a very sharp knife to score a cross around the fruit, then carefully remove the peel.

Place the peel in a pan with enough water to cover it and heat gently for an hour. Try to change the water a few times during the cooking, if you can, to remove any scum.

Strain, then use a teaspoon to remove the pith from inside the peel. This is the time either to cut the peel into strips or leave it as larger rustic pieces.

Weigh the peel, then return it to the pan and add the same weight of sugar. Add a little water, enough to dissolve the sugar, and place on a low heat for 45 minutes. Repeat the process with a second batch of syrup.

Lay the peel on a baking tray and leave to cool. Alternatively, you can place the tray in the oven at 80°C (175°F) or the lowest gas setting for an hour. Dip the dried peel in caster sugar and store in a sterilized jar (see page 53).

CHOCOLATE-DIPPED CANDIED PEEL

- 1 pink grapefruit
- 2 oranges
- 2 limes
- 1 lemon

Follow the numbered steps opposite and on this page, but when the peel is ready, melt some chocolate in a bowl over simmering water and dip half of each piece of candied peel into the sweet molten coating. Place on a rack or sheet of baking paper to set.

Full of lemony goodness

METHOD #18

FRUIT CURDS

A fruit curd is a dessert spread usually made with the juice and zest of citrus fruit (though other fruits can also be used), combined with egg yolks, sugar and sometimes butter. When the spread is cold it is a smooth and very flavoursome alternative to jam. It's lovely simply served on scones or fresh bread, but homemade fruit curds also make amazing fillings for tarts, cakes and the classic lemon meringue pie.

PREPARING THE FRUIT

Chilled citrus fruits are easier to grate or zest, but the zest dries out quickly, so cover it with clingfilm until you need it. Room-temperature citrus fruits will yield more juice – roll them on a work surface prior to squeezing to get the maximum amount of juice out of them.

MAKING THE CURD

There are two ways to make fruit curd: you can add the butter to the eggs (see below) or vice versa (see numbered steps opposite). Both methods work equally well, so we've included both.

Put the eggs, sugar, zest and juice into a heatproof bowl and whisk until well combined. Place the bowl over a pan of simmering water, making sure the water does not touch the bottom of the bowl, and stir continuously until the mixture becomes thick. Remove from the heat. If you want perfectly smooth curd, pass the mixture through a sieve to remove the zest. Stir in the butter. Pour into a sterilized jar (see page 53) and seal immediately. Keep in the fridge.

STORING

As fresh eggs are used, these fruit curds do not keep as long as jams and therefore tend to be made in relatively small batches. Fruit curd will keep for about 4 weeks in the fridge.

ORANGE CURD

Makes 500g (1lb)

- 3 oranges
- 3 eggs, well beaten
- 150g (5oz) caster sugar
- 75g (3oz) unsalted butter, cut into small cubes

Zest 1 of the oranges, then squeeze the juice from all 3 of them and put through a sieve.

To make the curd, follow the numbered steps opposite.

HOW TO MAKE LEMON CURD

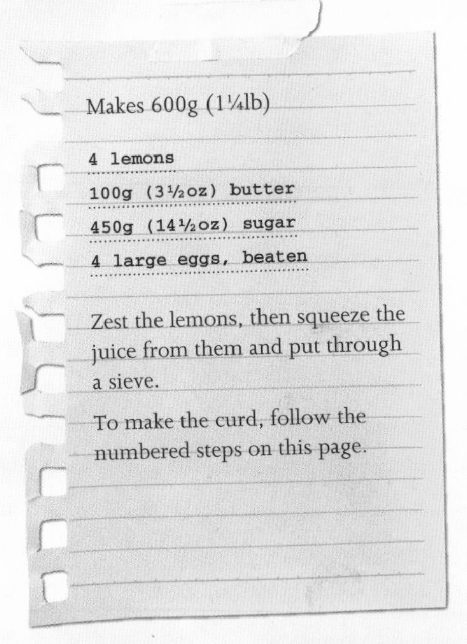

Makes 600g (1¼lb)

4 lemons

100g (3½oz) butter

450g (14½oz) sugar

4 large eggs, beaten

Zest the lemons, then squeeze the juice from them and put through a sieve.

To make the curd, follow the numbered steps on this page.

Put the lemon zest and juice, butter and sugar into a heatproof bowl over a pan of simmering water. Gradually heat, whisking until the butter melts, then take off the heat.

Beat the eggs well and gradually drizzle them into the lemon mixture, whisking all the time. Return to the heat and continue to whisk as the mixture heats up.

Cook for about 10 minutes, until the curd thickens, then pour into sterilized jars (see page 53). Seal and, when cool, store in the fridge.

This sharp-flavoured cake has an underlying sweetness. Feel free to make two cakes and sandwich them with lemon curd for a seriously tangy surprise.

SERVES 8

icing sugar, for dusting
2 tablespoons lemon curd
100g (3½oz) caster sugar
100g (3½oz) unsalted butter, softened
150g (5oz) self-raising flour
2 large eggs (at room temperature), beaten
zest of 1 lemon

FOR THE LEMON DRIZZLE

juice of 2 lemons
50g caster sugar

TO SERVE

crème fraîche
lemon curd (see page 117)

LEMON CURD CAKE

Preheat the oven to 170°C (325°F), Gas Mark 3. Grease a shallow 20cm (8 inch) cake tin and dust it with icing sugar.

Beat the lemon curd, sugar and butter in a large bowl. Sift in the flour while continuing to beat. Add the eggs and beat again, and finally add the lemon zest.

Pour the mixture into the cake tin and bake for 35 minutes, then allow to cool in the tin for 10 minutes, until the cake shrinks slightly away from the sides. Transfer to a wire rack and leave to cool.

Mix the lemon drizzle ingredients together and drizzle over the cooled cake.

Serve slices of the cake with a spoonful each of crème fraîche and lemon curd.

METHOD #19

FRUIT CHEESES

you have to try this

Fruit cheeses are made from a stiff fruit purée and are a great way of preserving fruit that has a lot of pips or stones as the pulp is passed through a sieve. The method works best with fruits that have a high pectin level (see pages 92–3). The basic process is simple: equal amounts of fruit pulp and sugar are heated gently for about an hour until the sugar has dissolved and the mixture thickened. Fruit cheeses are often spreadable and go really well with cheese and cold meats. They should be left for up to 2 months to mature and will keep for 4 months if stored in a sealed jar.

MEMBRILLO – QUINCE CHEESE

This is far and away our favourite fruit cheese. *Membrillo* is the Spanish word for 'quince', and in Spain they have been making a cheese with it in a similar fashion for hundreds of years. The golden fruits make a delicious sweet condiment. Quinces are particularly high in pectin, so they set very well, and the gorgeous coral pink colour looks great on a cheese board!

Makes 750g (1½lb)

- 1.5kg (3lb) quinces
- 1 vanilla pod
- caster sugar

Peel and core the quinces, then cut them into pieces and put into a large pan. Cover them with water and add the vanilla pod. Put a lid on the pan and boil for 40 minutes. When all the quince pieces are soft, remove from the pan with a slotted spoon and place in a bowl. Discard the quince syrup.

Weigh the quince pulp and return it to the pan. Add an equal weight of sugar and place on a low heat, stirring gently until the sugar dissolves. Keep cooking for a further hour, until the quince has thickened into a rich coral pink paste.

Preheat the oven to 50°C (120°F) or its lowest gas setting. Grease a baking tray and line it with baking paper. Pour the paste into the prepared tray and smooth it out to the edges. Place in the oven for 1 hour – this will help the cheese to set more quickly. When cool, slice into small portions and store in the fridge.

HOW TO MAKE A FRUIT CHEESE

Chop your chosen fruit, place it in a large, thick-bottomed pan and cover with water. Bring to the boil, then reduce the heat and simmer for 30–40 minutes, or until the fruit has softened.

Strain the liquid out of the fruit, then weigh the pulp. Put the pulp back into the pan and add the same weight of sugar. Place on a low heat and stir until the sugar has dissolved.

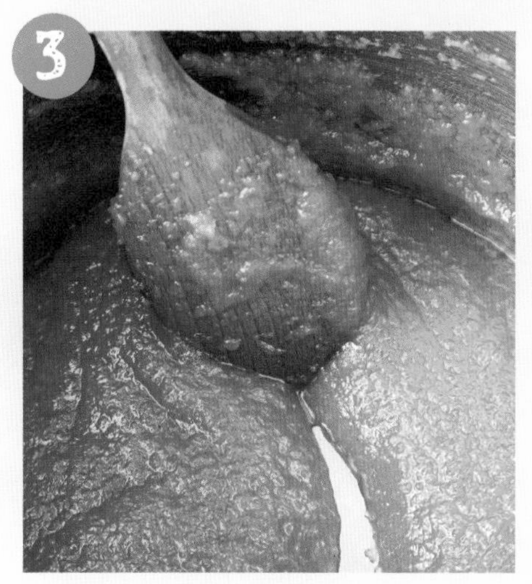

Continue to stir for an hour, still over a low heat, until the pulp thickens. To test for readiness, drag a spoon through the mixture: if it leaves a visible line, it is good to store.

Pour the fruit cheese into a jelly mould (or small bowls greased with glycerine if you like, so it's easy to serve later) and cover the surface with a circle of greaseproof paper. Leave for a week to mature before enjoying.

BOTTLES

INTRODUCTION TO

BOTTLES

Bottling is a very versatile preserving technique, and shows off the breadth of products that can be made at home beautifully. You can bottle everything from refreshing homemade beer or cider to herbal tinctures that can help relieve minor ailments such as coughs, colds or indigestion. Vegetable ketchups and fruit cordials, made when the ingredients are plentiful and at their best, can preserve those fleeting, seasonal flavours for enjoyment throughout the year. The only thing these diverse products have in common is that they are stored in bottles with sealable stoppers, but all of them will make useful – and very decorative – additions to your kitchen and store-cupboard shelves.

GUIDELINES

- Make sure your bottles and any other pieces of equipment you use are clean – wash them with boiling water or use cleaning and sterilizing solutions.
- The seals on the bottle tops must be in good condition. For flip tops this is easy to confirm, but when reusing older bottles, you must make a proper inspection.
- Always label your bottle with its name, the date of bottling and any other notes (if you have deviated from the recipe, for example).

CORDIALS

One old dictionary definition of a cordial is 'a comforting or pleasant-tasting medicine'. These days we think of a cordial not as something medicinal but as a sweet, non-alcoholic, usually fruit-flavoured viscous liquid that is diluted with water to make a refreshing drink. Our cordials are created with the best-quality fruits to produce a drink that can be enjoyed by all the family. It is worth remembering that there is more to these soft drinks than just adding water and ice: you can try adding them to hot water or apple juice and infusing them with spices to make a delicously warming drink, perfect on a cold winter's evening. Or in the summer, you could serve chilled dry white wine with a dash of fruit cordial for a fine aperitif. Raspberry, blackberry or peach are particularly good served in this manner.

WINES & CIDERS

The natural yeasts and sugars that are present in fruits and vegetables should be celebrated, and the person who discovered, thousands of years ago, that yeast turns those sugars into alcohol and carbon dioxide did us all a huge favour. It is extremely easy to make your own alcoholic drinks, be it wine, cider or even something a little stronger, though it must be said that it is more of an issue to make them well. You might not produce an exceptional wine first time round, but those early attempts are special, so enjoy every glass. Start with a small batch, take careful notes about what you do and how you like the results, and refine the process when you try again.

Remove the berries by hand or use a fork

Collect ripe grapes on a dry day and juice them as soon as possible

When it comes to making wine, most people will immediately think of grapes, but you can make wine from almost anything – rosehips, nettles and parsnips, or even pea pods.

Over recent years we have been making more cider, as there is an abundance of apples in our neighbourhood. It is quite a task, and we tend to get friends together so that all can share in the fun and the spoils. Some of our cider has been very potent, so we are quite careful not to drink too much. Our first batch of cider (about 90 litres/20 gallons) was extremely dry, so we turned it into spiced cider by adding lots of spices and sugar. Delicious!

SAUCES & KETCHUPS

A dash of hot homemade chilli sauce or a dollop of creamed horseradish will liven up many a dish, and it's always good to have a bottle of fresh pesto on stand-by in the fridge for a quick week-night supper. Ketchups are sweet, tangy condiments typically made with vegetables, vinegar, sugar and spices. Tomato ketchup is probably the most popular variety available today, but many other versions exist, from anchovy- and mushroom-based sauces to the banana ketchup popular in the Philippines.

TO MAKE TOMATO KETCHUP

Each brand of tomato ketchup has its own distinctive flavour, so when you make it at home you can expect to produce your own unique variation. Tomato ketchup is nothing new – the recipe below dates back more than 200 years and is quite salty, because salt acts as a preservative.

- Harvest ripe tomatoes on a dry day and squeeze them into a pulp with your hands.
- Add 500g (8oz) of salt per 100 tomatoes and boil for 2 hours, stirring occasionally to prevent the tomatoes from burning.
- Pass the tomatoes through a fine sieve to remove the skin and seeds.
- Add mace, nutmeg, allspice, cloves, cinnamon, ginger and pepper to taste.
- Boil slowly until the sauce is thick.
- Pour into sterilized bottles (see page 53).

According to the recipe, this ketchup would keep for 2 or 3 years, but then again, they did not have sell-by dates 200 years ago.

TINCTURES

A tincture is a herbal extract. The active ingredients of herbs and other plants are dissolved in alcohol – vodka works very well – or in a mixture of alcohol and water. All over the world, herbs have long been prized for their medicinal qualities as well as for their flavours. Purple coneflower – also known as echinacea – is believed to stimulate the immune system, relieve pain and reduce inflammation, and was used by Native Americans for hundreds of years. Chamomile, valerian and lavender are all thought to have a calming effect and are used as remedies for sleep disorders and anxiety; peppermint will soothe an upset stomach and aid digestion; and elderberry extract is used to treat coughs and colds.

The very act of soaking herbs, or combinations of herbs, in alcohol and pressing the goodness out of them can feel therapeutic in itself. In fact, making tinctures can encourage you to take some responsibility for looking after yourself. That said, we are not advocating these tinctures as alternatives to conventional medicine, just something to complement it. Not all tinctures are safe for oral consumption, so make sure to do your research first.

METHOD #20

CORDIALS

If you are looking for a way to preserve fruit flavours, making your own cordials is a very good move. Not only do they retain lots of flavour, but they also keep fresh in sealed bottles for a year. We use cordials diluted with water as a refreshing drink, or in alcoholic drinks as a fancy mixer. Another great use for cordials is to drizzle them over fruit salad or ripple them through ice cream. Mint, lime, orange and pomegranate are some of our favourites.

MAKING CORDIALS

To make a cordial you need to simmer your fruit with some water to extract all the flavour. Once the fruit has softened, strain and measure the liquid before returning it to the pan and adding 350g (11½oz) of sugar per 500ml (17fl oz) of juice. Put over a low heat until the sugar has dissolved, then pour into sterilized bottles (see page 53) and seal.

ELDERFLOWER CORDIAL

Makes 1 litre (1¾ pints)

1.5kg (3lb) sugar
15 elderflower heads, shaken to remove any insects
2 lemons, zested and sliced
1 teaspoon citric acid

Put 1 litre (1¾ pints) water into a pan and bring to the boil. Add the sugar and stir until dissolved. Pour into a large bowl and add the elderflowers, lemons and citric acid. Stir, cover with a tea towel and leave overnight. The next morning, strain through a muslin-lined sieve. Pour into sterilized bottles (see page 53) and store in the fridge or a cool cupboard.

ELDERBERRY & BLACKBERRY CORDIAL

Makes 1.5 litres (2½ pints)

250g (8oz) elderberries
250g (8oz) blackberries
sugar
juice of 1 lemon
1 cinnamon stick
1 teaspoon citric acid [optional

Put the berries into a pan with 150ml (5fl oz) of water and set over a low heat for 5 minutes, or until juice runs from the fruit. Cool, then mash with the back of a spoon. Strain through a jelly bag or a sieve lined with muslin, and measure the juice. Put it back into the pan and add an equal quantity of sugar. Add the lemon juice and cinnamon stick, bring to the boil and boil for 2 minutes. Skim off any scum. Pour into sterilized bottles (see page 53) and seal.

HOW TO MAKE A CORDIAL

Place your fruit in a pan with some water and simmer over a low heat until soft. Berries need only a little water, but firm fruits need 200ml (7fl oz) to 500g (1lb) of fruit.

Use a muslin-lined sieve or a jelly bag to strain the juicy fruit. Press it through gently with a ladle or the back of a wooden spoon.

Measure the juice and add sugar in the ratio of 350g (11½oz) of sugar to 500ml (17fl oz) of juice. Put back on a low heat, stirring to dissolve the sugar, then bring to the boil.

Skim off any scum, then pour the cordial into sterilized glass bottles, using a sterilized funnel (see page 53). Store your bottled cordial in the fridge for 1–2 months.

Elderflower syrups, cordials and liqueurs are flavoured with the scents of the summer. When pears are abundant in the autumn, this pudding is a reminder of those days. The poached pears are lovely with shortbread.

SERVES 4

4 Conference or other firm pears

250ml (8fl oz) elderflower cordial (see page 128)

250ml (8fl oz) water

juice and zest of ½ a lemon

4 mint sprigs, to decorate

FOR THE SYLLABUB

250ml (8fl oz) double cream

3 tablespoons elderflower liqueur

POACHED PEARS & ELDERFLOWERS WITH SYLLABUB

Peel the pears, leaving the stalks on, and position them snugly in a saucepan. Cover with the cordial, water and the lemon juice, cover with a lid and simmer for about 30 minutes, until the pears are soft. Turn them a couple of times during cooking, but be careful not to dent them.

When the pears are ready, transfer them to a plate and keep them in a warm place. Boil the liquor to reduce it to a thin syrup.

To make the syllabub, whip the cream to soft peak stage and fold in the elderflower liqueur. Put into a small bowl and drizzle a little of the pear syrup over the top.

Place each pear on a plate, glaze it with a little syrup, top with a mint sprig and serve with a spoonful of syllabub.

A snug fit

If you grow your own tomatoes, there is a very good chance you will end up with a glut at some stage. This spicy drink is a great way to use them. Some tomatoes are sweeter than others, so always use the ripest you can find.

SERVES 4

FOR THE TOMATO JUICE

2kg (4lb) very ripe red tomatoes

6 stalks of celery

½ an onion

2 tablespoons sugar (or to taste)

3 tablespoons red wine vinegar

1 teaspoon salt

a good pinch of white pepper

FOR EACH DRINK

1 measure of vodka

a dash of Worcestershire sauce

a slice of lemon

a celery stick

BLOODY MARY

To make the tomato juice, core the tomatoes, then roughly chop them with the celery and onion. Put all the ingredients into a large non-reactive pan (stainless steel, not aluminium). Bring to a simmer and cook, uncovered, until the mixture is soft – about 25 minutes – then force the mixture through a sieve. Let the juice cool completely, then store in the fridge – it will keep for about a week. To keep it longer, return the sieved juice to the pan, bring it to the boil and boil for 2–3 minutes. Pour into sterilized bottles (see page 53) and seal tightly. This will make about 1.5 litres (2½ pints) of tomato juice.

To make a Bloody Mary, pour the vodka into a tall glass, top up with the tomato juice, and add the Worcestershire sauce. Add a slice of lemon and stir with a celery stick.

A bottle of summer flavour

No air miles for this wine

METHOD #21

WINES

What could be better than relaxing in the evening with a glass of homemade wine? There is not much in the world that can rival the feeling of absolute satisfaction, pride and subsequent relaxation. For a small investment it is easy to make a great deal of drinkable wine; whether it turns out to be fine wine, however, very much depends on the care and patience you put into it. You also have the chance to try out some exotic recipes that are not available at any wine merchants we know of, with relatively little effort. Wine-making can be approached as a scientific experiment with exact measurements under meticulous conditions or, as we do, much in the spirit of wine itself, as a vibrant and fun hobby. We are not too bound to strict rules and prefer to let it evolve naturally. Having said that, there are a few basics that help achieve a fine wine.

GUIDELINES

- Bottles and all other apparatus must be sterilized with boiling water or sterilizing solutions. (Campden tablets can be used for this.)
- Wines need sugar, yeast, acid and tannin. The amounts needed will depend on the recipe you use.
- We follow the traditional advice about quantities of sugar for wine-making. This states that for every 4.5 litres (8 pints) of grape juice add: 1kg (2lb) of sugar for a dry wine; 1.25kg (2.5lb) of sugar for a medium wine; and 1.36kg (2.75lb) of sugar for a sweet wine.
- Yeast is naturally present in the air all around us, and also in the 'bloom' of fruit. You can rely on that to do the job of fermentation. However, we use good-quality wine yeast (not baker's yeast) for making our wines. It is available in granulated or liquid form and is best bought from a brewing shop or online.
- The crispness of a wine depends upon its level of acidity. You may need to add more acid to wines made from ingredients that aren't very acidic, such as flowers or grain. You can buy citric acid from brewing shops or old-fashioned chemists, but we tend to use lemon juice, which works fairly well.
- Tannins comes from the skin of the grape and give the wine that 'dry mouth' feel. You may need to add extra, depending on the type of wine being made. Tannin can be bought online or from a brewing shop.
- To make successful wine you need to keep the temperature favourable to the vital yeasts. During the first fermentation, aim for 24°C (75°F), and avoid it rising to over 27°C (80°F), when the yeast will start to

die, or under 21°C (70°F), when the yeast will be too cold. A warm place, such as an airing cupboard, is ideal, or you can invest in a heating belt.

MAKING WINE FROM GRAPES

The first step is to press your grapes any way you can. The traditional way of doing this is fun but involves getting your socks off and giving your feet a very good clean. You then squish the grapes under your feet and between your toes until they are reduced to pulp and swills of grape juice. Don't stop until you have squeezed every last bit out of the fruit. Alternatively, use a press: put in a generous layer of grapes wrapped in mesh cloth and then add other layers on top of it until you have used all of your grapes or run out of space.

FERMENTING THE GRAPE JUICE

Once you've pressed the grapes, strain the liquid through a sieve lined with muslin, then pour it into a fermentation jar or demijohn using a funnel. Put a fermentation lock in the top of the jar – this will allow gases to escape but stop outside air contaminating the contents.

There are two stages to fermenting your own wines. The first is the frothy, active stage, when the yeasts are multiplying and need air around them. Leave the jar or demijohn three-quarters full for this stage, which will last for about 10 days. It is complete when the froth dies back. For the second stage, top the jar up with water and place an airlock or fermentation lock on the top. This allows the gases produced during fermentation to escape while keeping air out, thereby avoiding oxidation. It also prevents bacteria and other bugs from ruining your wine.

RACKING THE WINE

Your wine is ready to be racked when the sediment settles and the wine looks clear – it usually takes 8–10 weeks to reach this stage. Racking essentially means siphoning your wine off from above the yeasts (also known as the lees) that have settled at the bottom, transferring it from one container to another. It is important to rack your wine so that the lees don't spoil its flavour. To do the siphoning, use a simple length of flexible, clear plastic tubing. A month after racking your wine the process should be repeated, and, if you have the patience, do it again 3 weeks later. Before the final racking leave your wine somewhere cold to hasten the settling of any leftover sediment.

BOTTLING AND CORKING

When the wine is 6 months old and the fermentation process is complete, you can bottle and cork it. Sterilize your bottles (see page 53), choosing ones made of dark-coloured glass to preseve the colour of the wine. Use a siphon to pour your wine into bottles, leaving 4–5cm at the top for the cork. Drive in the corks using a corking gun or a wooden mallet. Lable the bottles clearly and store them on their sides at a temperature of about 13°C (55°F) for a full year.

NOW TRY: PEA POD WINE

To make pea pod wine, place 2kg (4lb) of pea pods in a large pan with 5 litres (1 gallon) of water and bring to the boil. Reduce to a simmer for 20 minutes, then remove from the heat, strain off the liquid and add 1kg (2lb) of sugar, 1 tablespoon of dried yeast, 2 teabags and a sliced lemon. Cover with a tea towel and leave for 3–4 days, then strain and put into a demijohn with an airlock for 4 weeks. Rack, bottle and cork as above.

HOW TO MAKE WINE FROM GRAPES

First, clean your feet thoroughly, then put your grapes into a container large enough to stand in.

Press the grapes under your feet until they are reduced to pulp and you have extracted all the juice.

Put the resulting juice through a sieve lined with a fine muslin to remove any sediment.

Transfer the juice to a sterilized demijohn (see page 53) and allow natural yeasts to do the fermenting, or use good-quality wine yeast (see page 134).

Line a large colander with open-weave muslin, place into over a large container and pour in the grape juice and pulp.

Twist the muslin tightly and squeeze the juice into the container below. Keep squeezing until you have got as much juice as possible.

Add an airlock to your demijohn to keep oxygen out of the fermentation process. The carbon dioxide produced pushes out any oxygen and bubbles away until the sugars in the fruit change to alcohol.

Use plastic tubing to rack your wine, siphoning the wine from above the yeasts that have settled at the bottom into another sterilized demijohn. Store for another month, then rack again. Bottle and cork as described on page 135.

Pea pod wine is a delicious drink, but you can cook with it too. This recipe features a pea-pod wine and mint reduction that is the perfect accompaniment for rolled lamb.

SERVES 4

- a knob of butter
- 2 shallots, finely chopped
- 1 garlic clove, finely chopped
- 150g (5oz) peas
- 1 tablespoon chopped fresh mint
- 1 tablespoon olive oil
- salt and freshly ground black pepper
- 75g (3oz) breadcrumbs
- 1 breast of lamb, weighing about 400g (13oz)

FOR THE REDUCTION

- 150ml (5fl oz) pea pod wine (see page 135)
- 4-6 fresh mint leaves, finely chopped
- 1 tablespoon white wine vinegar
- 1 tablespoon sugar

TO SERVE

- roast potatoes
- braised cabbage and pancetta

ROLLED LAMB WITH PEA POD WINE

Preheat the oven to 160°C (325°F), Gas Mark 3. Melt the butter in a pan, add the shallots and garlic, and cook gently until softened. Add the peas, stir for a minute and remove from the heat. Add the mint, olive oil and a pinch of salt and pepper. Put this mixture into a blender with the breadcrumbs and blitz for a few seconds.

Lay the breast of lamb out flat and spread the pea stuffing over it. Roll up and tie with butcher's string, then put the lamb into a roasting tray and cook in the oven for 1½ hours. Meanwhile, make the reduction. Put the wine into a small pan and simmer for 10–15 minutes, until reduced by half. Add the mint leaves, vinegar and sugar and simmer for a further 3–4 minutes. Strain and set aside.

Remove the lamb from the oven, cover with tin foil and allow to rest for 10 minutes.

Warm the reduction. Slice the lamb, pour over the reduction and serve with roasted potatoes and braised cabbage and pancetta.

No bubbles, all flavour

METHOD #22

CIDERS

Cider is the fermented juice of apples that have been crushed and then pressed. Every year we make cider with apples from our small orchard, and we also help out our neighbours by taking some of their unwanted apples in exchange for some bottles of cider. On the whole, apples don't last very long unless they are picked when ripe and are not at all bruised or damaged. Making cider means that you can successfully preserve huge quantities of fruit and save lots of money on alcohol by doing so. Cider-making is also a really social activity that can be a great way to pass an autumn day.

CHOOSING APPLES

Any variety of apple can be used to make cider, and even dodgy-looking windfall apples can be turned into good scrumpy. Unless you are growing specific cider-making apples, such as Langworthy, Foxwhelp or Crimson King, which can be brewed on their own, the trick to making a tasty cider is to use a mixture of apples. Use one-third each of bitter-sweet, sweet and sharp apples. A balanced cider should have equal proportions of bitter-sweet apples that are low in acid but high in tannins, such as Dabinett, Somerset Red and Yarlington Mill, sweet dessert apples that have a medium acidity and are low in tannin, such as Cox's Orange Pippin, Golden Delicious and Cornish Gilliflower, and finally some sharper apples with higher acidity, such as Royal Russets or Herefordshire Costards.

PREPARING APPLES

Before you crush your ripe apples, leave them in a heap for 2 or 3 days to soften. Alternatively, pick up your windfall apples and they will already be very juicy. Avoid using apples that have serious pest problems, but most apples will generally be usable even if they don't look perfect. These apples will contain some of the natural yeasts called 'bloom' that will ferment later in the process. Apples vary greatly in the amount of juice they create, but on average we would expect to need about 8kg (17½lb) of apples to make a gallon (5 litres) of cider.

CRUSHING APPLES

Make sure that all your equipment is clean. An electric apple crusher or masticator (see page 142)will enable you to make large volumes of cider quickly and easily, thus adding to the overall enjoyment of the age-old tradition.

Another simple way to crush the apples is to put them in a wooden box and use a spade as a chopper. It needs to be a strong wooden box and the apples should be near the top and closely packed together so that they don't easily move. Don't pack it so full that they will jump out and spill on to the floor. Use a sharp clean spade to chop the apples into small pieces.

PRESSING

Spread your crushed apples out evenly in layers in your press, covering each layer with a sheet of fabric mesh or muslin, and press them so that the juice runs straight into a sterilized demijohn or large fermenting bin. This is the time to savour a moment of magic and try the cloudy apple juice before it transforms into a golden alcoholic liquid. If you place a large jug under the press, you can collect some delicious fresh apple juice, which will keep in the fridge for a few days.

When you have squeezed out all the juice from the first batch, unwrap the leftover pulp, also known as the 'cheese', and feed it to one of those famous waste-disposal experts, the pigs or the compost bin. Then refill your press and repeat the steps until you have pressed all your apples.

FERMENTING

Once you have your fresh apple juice all stored in demijohns or a large fermenting bin, it's time for the yeast to work its magic. We allow the natural yeasts from the apple skins to do the fermenting, but if you want to ensure success, add a few teaspoons of wine-maker's yeast to the mix. You can also add sugar or syrup to the cider if you prefer a sweeter brew.

BOTTLING

Once fermentation has ceased, any time from 10 days to a month, you can transfer the cider straight into bottles for storage using a siphon (see page 137, step 8). At this stage you will often find that a secondary fermentation takes place in the bottle. Cider is ually drunk within a year, but it can be kept for longer. We once bottled some apple juice after trying to pasteurize it and a couple of months later we were greatly surprised to find it had turned into a bubbly alcoholic drink similar to the fizzy cider produced in Normandy – of course we didn't let on about the mistake to our guests, who loved it!

NOW TRY: PERRY

Wherever the soil is good for growing apple trees, pears will also flourish. To make perry or pear cider, follow exactly the same method as for apple cider, but use pears instead. Perry tends to be much sweeter than cider, so you don't need to add any extra sugar during the fermenting process.

HOW TO MAKE CIDER

Discard any rotten fruit, then chop up your apples using a masticator.

Spoon a layer of the chopped fruit on to a sheet of fabric mesh in your press. The layer should be about 5cm (2 inches) deep. Fold in the edges of the fabric to make a parcel.

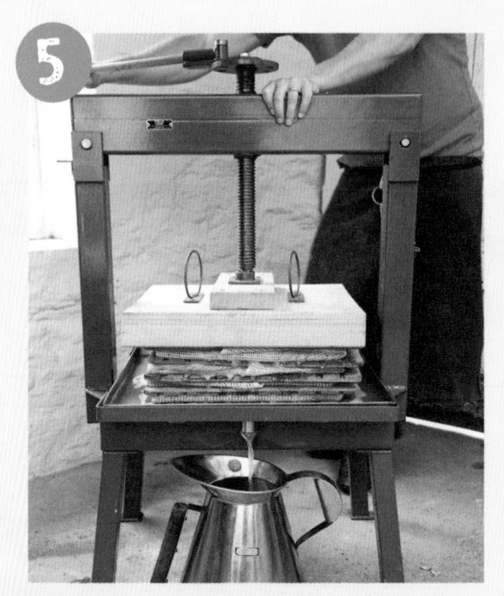

Screw down the press until as much juice as possible has been extracted. If your press is not screwed to the floor, you might need two people to do this: one to hold the press, the other to turn the lever.

Transfer the pressed juice to sterilized demijohns or brewing barrels.

Place a wooden rack on top of the folded fabric. Repeat steps 2 and 3 until you have filled the press, finishing with a wooden rack.

Place a heavy wooden block on top of the final wooden rack, and place a large jug under the spigot.

Add an airlock that allow the gases produced during fermentation to escape but stops air getting in to oxidize the cider.

When fermentation is completed and the yeast and sediments have settled at the bottom of the jar, siphon off the good cider into a new demijohn or bottles.

This mussels recipe is our version of moules marinière but with a homemade twist. The Armagnac will add a bit of intensity and really bring out the flavour of the cider.

SERVES 4

2kg (4lb) mussels, cleaned and beards removed

20g (¾ oz) butter

140g (4½ oz) dry-cured streaky bacon, diced

1 bay leaf

2 or 3 sprigs of fresh thyme

3 shallots, halved

50ml (2fl oz) Armagnac (optional)

150ml (5fl oz) cider

200ml (7fl oz) crème fraîche

salt and freshly ground black pepper

MUSSELS WITH CIDER, ARMAGNAC & THYME

Check over the mussels, discarding any that are open or that don't shut when you tap them.

Melt the butter in a large pan, add the bacon and fry for 3–4 minutes. Add the bay leaf, thyme and shallots and cook for a few more minutes. Then add the Armagnac, if using, the mussels and the cider and cook on a high heat for 5 minutes, stirring occasionally.

Remove the mussels from the pan with a slotted spoon when they have all opened. Discard any that are still shut. Simmer the sauce for a few minutes more and stir in the crème fraîche. Season well with salt and pepper and serve.

All lined up and ready to use

METHOD #23

SAUCES & KETCHUPS

Sauces and ketchups are useful as condiments and can add concentrated flavour to all sorts of dishes. We also find that when it comes to using and preserving vegetables in bulk, they are one of the most effective ways to keep things fresh and tasty. Experiment using the ketchups in your cooking rather than eating them alongside your finished dish: the resulting flavour is much deeper.

PREPARATION

Start by washing or peeling your fruit and vegetables as necessary. Those that are soft enough can then be puréed raw in a blender. Others can be cooked slowly on a low heat until soft, and then blended. Herbs and chillies often remain more potent if you blitz them straight into a sauce. Adding oil, vinegar or citrus juice will help to retain their fresh taste – the amount needed will vary from recipe to recipe.

ADDING FLAVOUR

Adding sweetness and spice is best done before you strain the sauce. Use a pestle and mortar to bash up the herbs and spices roughly, then stir them into the fruit pulp or reduced vegetables to infuse with flavour.

STRAINING

For a thin ketchup, strain through a very fine sieve or a piece of muslin, but for thicker condiments don't worry about some lumps. The key thing is to remove any fragrant whole spices before bottling.

BOTTLING AND STORING

Pour the finished sauce into a sterilized glass bottle (see page 53) and store in the fridge if possible. Consume within 1 month.

GREAT SAUCES

- PESTO SAUCE - This classic combination of basil, olive oil, pine nuts and Parmesan cheese is amazing, but it's also worth experimenting with other ingredients, such as sage and walnuts, or parsley and pumpkin seeds. Pesto excellent for making quick pasta dishes at lunchtime.
- CHILLI SAUCE - The hotter the better! You can play around with the ingredients for chilli sauce, but a good combination is garlic, ginger, coriander, smoked chilli and vinegar.
- TOMATO KETCHUP - Thicker than other ketchups, this classic is great with loads of dishes. A homemade version really is hard to beat, and is a usually made with ripe tomatoes, salt, mace, nutmeg, allspice, cloves, cinnamon, ginger and pepper (see page 127).

HOW TO MAKE MUSHROOM KETCHUP

Makes 400ml (14fl oz)

2kg (4lb) portabello or field mushrooms, chopped
25g (1oz) salt
1 small onion, diced
2 cloves
1 teaspoon black peppercorns
3-4 allspice berries
300ml (½ pint) white wine vinegar
1 tablespoon soy sauce
1 tablespoon brandy (optional)

If you end up with more than 400 ml (14fl oz) of ketchup, reduce it over a low heat until the flavour is more intense. It will keep for up to a year.

Put the mushrooms in a shallow dish and cover with salt. Leave for 24 hours, occasionally squashing them and mixing in the salt.

Combine all the other ingredients, except the brandy, in a pan and bring to the boil. Cover and simmer on a medium heat for 1 hour, until you have about 400ml (14fl oz) of liquid.

Pass the mixture through a sieve into a jug. Add the brandy, if using. Put a muslin-lined funnel into the neck of a sterilized bottle (see page 53) and pour in the ketchup. Seal tightly and keep in the fridge.

Horseradish and smoked mackerel go brilliantly together, and as canapés these stuffed choux balls are hard to beat. To make the filling even more quickly, you can use shop-bought mayonnaise and horseradish sauce, but we like to make our own. Multiply the recipe, and enjoy the extra sauce with sandwiches.

MAKES ABOUT 12

- 2.5cm (1 inch) horseradish root (more or less, according to how hot you like it and how thick your horseradish is)
- 100ml (3½ fl oz) mayonnaise
- zest of ½ a lemon
- 2 smoked mackerel fillets, skinned
- wedges of lemon, to serve

FOR THE CHOUX PASTRY

- 75g (3oz) butter
- 125ml (4fl oz) water
- 75g (3oz) plain flour, sifted
- 2 eggs, beaten

HORSERADISH & SMOKED MACKEREL CHOUX BALLS

Preheat the oven to 220°C (425°F), Gas Mark 7. Line a baking tray with non-stick baking paper or use a silicone tray.

First, make the choux pastry. Put the butter and water into a medium pan and heat until the butter has melted. Bring to the boil, then immediately take off the heat and add the flour in one go. Stir until the mixture pulls away from the sides of the pan to form a ball – don't over-stir at this stage. Allow the mixture to cool down a little, then add the eggs a little at a time, beating until the mixture forms a smooth dough.

Using 2 tablespoons, or a piping bag fitted with a large nozzle, put dollops of the dough on the baking tray, spacing them well apart. Bake for about 20 minutes, until golden brown. Transfer the balls to a wire rack and leave to cool.

Meanwhile, make the filling. Finely grate the horseradish and combine with the mayonnaise and lemon zest. Flake the smoked mackerel, then mash it into the horseradish mayonnaise with a fork. Set aside until needed.

When the choux balls are cool, use a small knife to make a slit in the side of each one and pop in a heaped teaspoon of the filling. Serve with wedges of lemon.

A little goes a long way

METHOD #24

ALCOHOL TINCTURES

It's amazing how many plants have healing qualities, and making tinctures from them is an age-old practice. In the past herb tinctures were household remedies for all sorts of ailments, and they still have their uses today, but never use a tincture if you're pregnant and always consult your doctor first. Alcohol extracts and dissolves the active substance from herbs at the same time as preserving them. Fresh herb tinctures will yield more than dried herb tinctures, so you should really consider growing your own herbs at home.

MAKING FRESH HERB TINCTURES

Tinctures at their most basic involve dissolving a herb or solid in vodka or ethanol. The minimum amount of alcohol required to inhibit the growth of unwanted organisms is 25%, so if you are making your own remedies at home, a typical, standard vodka is more than adequate for the task. Shake the mixture every day for 2 weeks to transfer the herb's qualities to the liquid. When ready, pour the liquid through a piece of muslin and squeeze out every drop.

STORING TINCTURES

Sterilize a small glass bottle fitted with a dropper (see page 53) and fill it with your tincture. Label and store in a dark place. Tinctures can last for years, but the potency will reduce over time.

USING TINCTURES

To use, simply dissolve a few drops in water and drink.

TUMMY TINCTURE

This remedy is ideal for those who suffer from indigestion. Take it after eating, or beforehand to stimulate your digestive enzymes and promote absorption of good nutrients.

- 100g (3½oz) fresh rosemary
- 90g (3¼oz) fresh bay leaves
- 400ml (14fl oz) vodka

To make the tincture, follow the numbered steps opposite.

HOW TO MAKE A HERBAL TINCTURE

Pick your herbs, wash well and allow them to dry. Chop the herbs finely.

Place the chopped herbs in a large jar and pour the vodka over them.

Close the lid tightly and shake the jar well. Continue to shake the jar daily for at least 2 weeks.

Strain the contents through a muslin-lined sieve into a jug, and squeeze out as much of the liquid as possible. Store the tincture in sterilized dark-coloured dropper bottles (see page 53).

Propolis is a sticky resin that bees collect and it has fantastic antiseptic properties. We have been making this drink to help with sore throats and infections since we first started keeping bees many years ago. You can use it in varying amounts, depending on how sore your throat is feeling. We find it incredibly effective.

FOR THE TINCTURE

10g (½oz) propolis

30ml (1¼fl oz) vodka

FOR THE DRINK

4–6 drops of propolis tincture

1–2 teaspoons honey

1 slice of lemon

4 cloves

a sprig of fresh thyme

1–2 fresh sage leaves

BEE-HEALTHY PROPOLIS DRINK

To make the tincture, place the propolis in a jar, add the vodka, and shake every day for 2–3 weeks, until all the propolis has dissolved. Strain the liquid through a coffee filter to remove any bits of bee barb that may have got mixed with the propolis. Transfer to a sterilized glass bottle (see page 53) and store in a dark place.

To make the propolis drink, simply put 6–8 drops of the tincture into a mug, add the honey, lemon, cloves, thyme and sage and top up with boiling water.

Soothes a sore throat

5

FREEZING

INTRODUCTION TO

FREEZING

Enzyme activity within food cells, or bacterial activity from outside, means that no food will keep indefinitely in its natural form: everything is subject to the processes of decay and deterioration. The idea behind all forms of preservation is either to inhibit the activity of disease-causing micro-organisms, or to kill them altogether. Keeping food in the refrigerator will slow down bacterial action, but freezing it can stop stop the action entirely, since frozen bacteria are completely inactive. This means you can enjoy your frozen produce for longer.

GUIDELINES

- Use the correct type of packaging or container and ensure that the food is tightly wrapped or sealed.
- As with all preserving, ensure the food is in perfect condition.
- Divide the food into small portions to ensure rapid freezing and that smaller ice crystals are formed – the defrosted food will be of better quality.
- Label items clearly, including their expiration date (see chart on page 159).
- Check that your freezer is operating at -18°C (0°F).
- Don't try to freeze too many unfrozen items at once.
- Leave space around any new items that have been put into the freezer so that cold air can circulate. Pack them more tightly once they are completely frozen.
- Restock your freezer regularly – a full freezer is more efficient than a half-empty one.
- Defrost items in the fridge.
- Visit the freezer before you go shopping – use your own top-quality produce rather than spending money at the supermarket.

FREEZING BASICS

While refrigerators have a minimal effect on the texture of food, you do have to be a little more aware of the consequences of freezing. The cells in food contain a great deal of water, and the majority of foods will freeze solid at -5°C (23°F). If it is done correctly, freezing has little adverse impact on the quality of raw cuts of meat and minimal effect on most vegetables. However, freezing can completely change the texture of some types of fruit, which can become soft and mushy when defrosted.

If the food is frozen slowly, the water will form large ice crystals, which can break through the walls of the cells and will affect the texture of the produce. When the food is subsequently thawed, much of the liquid can drain out of these damaged cells, taking with it some of the minerals and vitamins and destroying the food's unique taste and texture as a result. Rapid freezing causes smaller ice crystals to form, so the damage to the cells is minimized and, once thawed, the resultant food is much closer to its original state.

Invest in an ice cream maker - you'll enjoy it

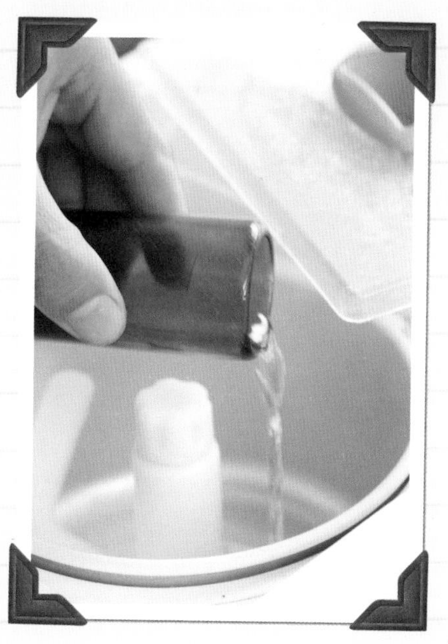

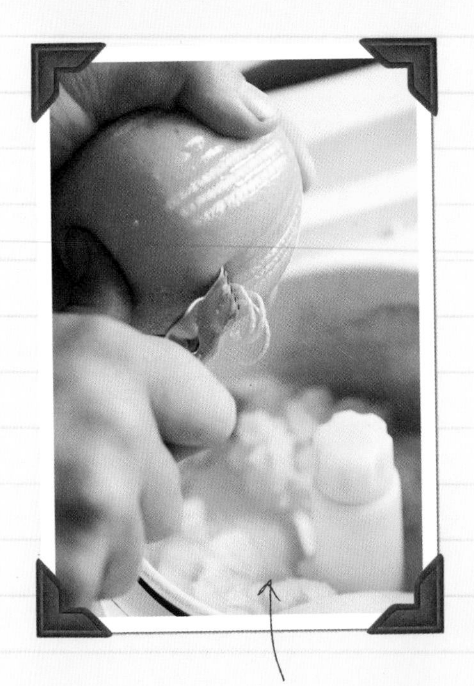

orange zest adds a lovely citrusy flavour

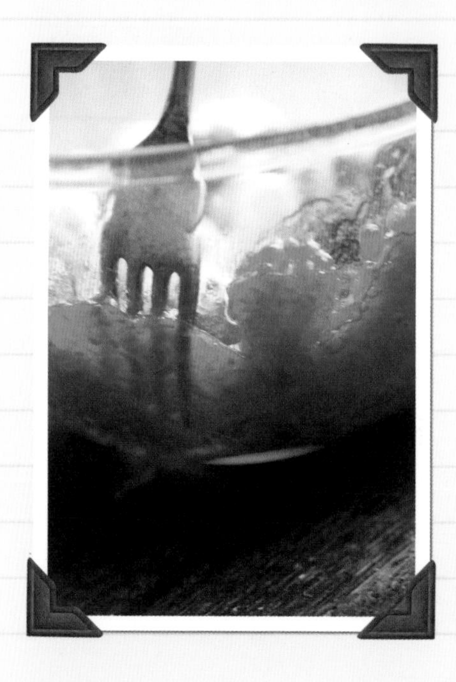

ORGANIZATION

It is quite common to fill your freezer with fresh produce, forget all about it and only discover it again months – if not years – later. Freezers need to be well organized, as it's not worth going to the trouble of processing your food if you are going to store it, lose it in the depths of your freezer and end up throwing it away. That is why we prefer an upright freezer to a chest freezer. The shelves and compartments make it that much simpler to stock effectively, to keep a clear overview of your supplies, and to quickly locate and remove the food you want to use.

PACKAGING

It is essential that the food going into the freezer should be tightly and carefully wrapped in moisture-proof or vapour-proof containers or materials. Expel as much air as possible from any packaging. Make sure that no air can enter, and that no moisture can escape or evaporate from the package. With a little care, anything frozen at its optimum freshness should emerge from the freezer in perfect condition.

If you pack and wrap your foods properly, you can avoid freezer burn, which leads to a loss of nutrients, flavour, colour, texture, taste, moisture and quality. Food that has been attacked by freezer burn loses moisture, is covered in frost, and greyish-white spots or patches develop on the surface of the food. Similar deterioration may occur if the food comes into contact with air. In most cases the burnt food is still edible – it is possible to cut away affected areas before or after the food has been defrosted, but the food will be dry and tough. Food can also be affected by freezer burn if it is stored for too long.

Clearly label your items: include the date on which they were frozen as well as the date by which they should be used (see chart below).

DEFROSTING

Having arrested bacteria growth in the freezer for the duration of the storage, we should not forget that it can take hold agian once the product is being thawed. As soon as food begins to defrost, the temperature rises, and, as moisture is present, bacteria will start to grow. If the food is thawed at a temperature above 40°F (5°C), bacteria will multiply rapidly in a very short time and can lead to food poisoning. For this reason, it is very important to defrost food safely, preferably overnight in the refrigerator.

STORAGE TIMES

If your freezer is set to an energy-efficient -18°C (0°F), most goods can be kept for up to 3 months. With a lower temperture setting of -30°C (-85°F), you can store some items for up to a year – use this chart as your maximum guideline.

	STORING TIME
RAW VEGETABLES	12 months
RAW MEAT	12 months
RAW WHITE FISH	6–12 months
RAW OILY FISH	4 months
UNCOOKED PASTRY	3–6 months
COOKED FOOD	2 months

Treat yourself to a scoopful

METHOD #25

ICE CREAMS

There is almost no limit to the flavourings you can add to ice cream to personalize it. Indeed, the bases for ice creams are varied as well, from lighter ones made with milk to those made with double cream. For us, however, the flavour and texture of a custard-based ice cream is very hard to beat – and a simple vanilla ice cream is one of our favourites. The easiest way to make ice cream is to use an ice cream maker, but we've provided instructions for making ice cream by hand as well.

HAND-MAKING ICE CREAM

Make a custard (see steps 1–4), then add your chosen flavouring. Put your liquid ice cream into a lidded plastic container and place in the freezer for about 2 hours, until the liquid freezes at the edges. Empty the contents into a bowl and whisk to break down the ice crystals. Return the ice cream to the container and place in the freezer for another couple of hours. Repeat these steps up to three times to get creamy ice cream.

VANILLA ICE CREAM

Makes 600ml (1 pint)

- 250ml (8fl oz) milk
- 1 vanilla pod, split in half
- 4 egg yolks
- 100g (3½oz) caster sugar
- 250ml (8fl oz) double cream

To make the ice cream, follow the numbered steps on this page and opposite.

HOW TO MAKE ICE CREAM

Put the milk and vanilla pod into a pan over a low heat. Bring to a simmer but do not allow it to boil. Turn off the heat and allow the milk to stand for 20 minutes, then remove the vanilla pod and scrape the seeds into the milk.

Put the egg yolks and sugar into a bowl and beat until creamy. Add 2 tablespoons of the flavoured milk to the eggs and beat in. Add 2 more tablespoons of milk and stir well.

Slowly add the rest of the milk, stirring continuously. Return the mixture to the pan and heat very gently, stirring until thick. Do not allow it to boil.

Remove from the heat and set aside. At this point you can add fruit or a liqueur of your choice, but remember that if you use too much alcohol, the ice cream will not set firm.

When cool, stir in the cream and either follow the instructions for your ice cream maker or finish the ice cream by hand (see page 160). Store in the freezer for up to 3 months.

This is our version of raspberry ripple. Adding herbs or spices to your ice creams can create some really unusual flavours. Here the addition of sage turns an old classic into something a little more sophisticated.

SERVES 8–10

2 large eggs

150g (5oz) caster sugar

500ml (17fl oz) double cream

2 fresh sage leaves, stems removed, finely shredded

300g (10oz) raspberries

1 tablespoon icing sugar

FOR THE SHORTBREAD

200g (7oz) plain flour

100g (3½ oz) unsalted butter

75g (3oz) icing sugar

2 fresh sage leaves, main stalk removed, finely chopped

RASPBERRY & SAGE ICE CREAM WITH SHORTBREAD

Whisk the eggs in a large bowl until light and frothy, then gradually add the caster sugar and whisk for 2 more minutes. Pour in the cream and whisk well. Once mixed, stir in the shredded sage leaves.

Pour the mixture into an ice cream maker and freeze to a soft scoop consistency according to the manufacturer's instructions. Alternatively, follow the instructions on page 160.

Meanwhile, place the raspberries and icing sugar in a bowl and crush them lightly with the back of a fork. When the ice cream has frozen to a soft scoop consistency, gently fold in the raspberries to create a rippled effect. Transfer to a plastic container and freeze until solid.

To make the shortbread, preheat the oven to 160°C (325°F), Gas Mark 3. Put the flour, butter and icing sugar into a food processor and whizz until the mixture looks like breadcrumbs. Take out the blade and mix in the chopped sage leaves.

Press the mixture into a small baking tray, about 20 x 15cm (8 x 6 inches). Cook in the oven for 20 minutes, until golden. Cut up the shortbread in the tray, then place on a rack to cool.

Serve the ice cream with a slice of shortbread.

Instantly refreshing

METHOD #26

SORBET

A sorbet is a semi-frozen mixture of fruit juice or purée and sugar syrup, and is softer and grainier than ice cream. We love to make them regularly as a great way to preserve that succulent quality and intense freshness that you get from a fruit when it's in season. It may sound a bit fancy, but cleansing your palate with a homemade seasonal sorbet is a great way to break up a meal -- alternatively serve up a decent scoop as a dessert and enjoy with fresh berries or a warm slice of cake.

CHOOSING INGREDIENTS

The key to a good sorbet is to use tasty ingredients. You can make a sorbet out of just about anything, but fruit in particular lends itself to this preserving method. You want the liquid to become an amalgamation of flavours.

STRAINING

Sorbets are often clean-tasting with a zing from the ice-cold crystals and intense fruit flavour. Straining your sorbet mixture through a fine sieve before it goes in the machine keeps it clear. If you want to add any extra texture, do this at the end.

ALCOHOL

Although alcohol doesn't freeze, it is often used when making a sorbet. Apart from adding flavour, it also makes the finished sorbet less grainy and smoother to eat. If you don't want the flavour to mess with your sorbet but you still want it to have a 'kick', vodka is the most discreet alcohol to use.

ORANGE & ROSEHIP SORBET

Makes 750ml (1¼ pints)

- 750g (1½ lb) rosehips
- 250ml (8fl oz) water
- 250g (8 oz) sugar
- 500ml (17fl oz) orange juice
- zest of 2 oranges
- 2 tablespoons vodka
- extra sugar, to taste

To make the sorbet, follow the numbered steps opposite.

HOW TO MAKE ORANGE & ROSEHIP SORBET

Skin or peel the fruit. Put in a pan with water, sugar, orange juice and zest, and simmer until soft and the juices have been released. Strain through a fine sieve.

Sweeten the solution with your extra sugar, if necessary, add spices or any further flavours, and finally allow to cool.

Add the vodka to the fruity sugar solution, or lace the ice cream maker with a generous shot before adding the mixture.

If using a machine, follow the manufacturer's instructions. If making by hand, see page 160. Store in the freezer for up to 3 months.

Savoury fruits, such as tomatoes, can really shine when you eat them in a different way, and this sorbet is delicious with the fresh salsa and toasted brioche beneath. It makes for an impressive starter or a special lunch.

SERVES 4

- 3 ripe avocados
- 400g (13oz) cooked white crab meat
- 1 fresh red chilli, finely chopped
- 1 tablespoon chopped fresh coriander
- 3 spring onions, finely chopped
- 1 teaspoon sesame seeds
- salt and freshly ground black pepper
- 4 slices of toasted brioche, to serve

FOR THE SORBET

- 100g (3½oz) sugar
- 100ml (3½ fl oz) water
- 2 tablespoons olive oil
- 1 onion, finely chopped
- 1 teaspoon orange zest
- a sprig of fresh thyme
- 1kg (2lb) tomatoes, seeds removed, finely chopped
- 6–8 fresh basil leaves
- 1 teaspoon salt

FOR THE DRESSING

- 1 teaspoon sesame oil
- 1 teaspoon crushed fresh ginger
- 1 teaspoon honey
- juice of 1 lime

CRAB SALSA WITH TOMATO SORBET

First make the sorbet. Put the sugar and water into a small pan and bring to the boil, then simmer until the volume has reduced by half. Set aside to cool.

Heat the olive oil in a pan and add the onion, orange zest and thyme. Cook until the onions are soft but not coloured. Add half the tomatoes and all the basil, and cook for 5–10 minutes, or until the water has evaporated and the mixture has thickened. Remove from the heat, discard the thyme sprig, and blend the mixture with the rest of the tomatoes, the salt and the reserved sugar syrup until smooth. Pass through a fine sieve into a bowl. Transfer it to an ice cream maker for 10 minutes of stirring and chilling, then put it into the freezer for 2–3 hours. If making by hand, follow the instructions on page 160.

To make the salsa, spoon the avocado flesh into a mixing bowl. Add the crab, chilli, coriander, spring onions and sesame seeds, season with salt and pepper and mix gently together. Whisk the dressing ingredients together in a bowl.

Place a slice of toasted brioche on each plate, add some spoonfuls of salsa, drizzle with the dressing and top with a scoop of tomato sorbet.

The taste relationship between orange and chocolate is dangerously addictive. To give the classic combination a fiery edge, we use chilli chocolate in the brownies, but any good chocolate will do.

SERVES 4

- 150g (5oz) plain flour
- 1 teaspoon baking powder
- 200g (7oz) dark chocolate, broken into squares
- 250g (8oz) butter, diced
- 300g (10oz) caster sugar
- 4 large eggs, beaten
- 100g (3½oz) walnuts
- 150g (5oz) chilli chocolate, broken into small pieces

FOR THE SORBET

- 200g (7oz) sugar
- 200ml (7fl oz) water
- juice from 6 large oranges (approx. 400ml/14fl oz)
- 1 star anise
- zest of 1 orange, cut into small strips

CHILLI CHOCOLATE BROWNIES WITH ORANGE SORBET

First make the sorbet. Put the sugar and water into a medium pan and bring to the boil, then reduce the heat and simmer for 10–15 minutes, or until the volume has reduced by half. Add the orange juice and star anise, bring back to the boil, then turn off the heat and allow to cool. When cool, strain through a fine sieve and place in your ice cream machine with the orange zest. Freeze according to the manufacturer's instructions. Alternatively, make by hand as described on page 160.

Preheat the oven to 180°C (350°F), Gas Mark 4. Line a 45 x 25 x 5cm (18 x 10 x 2 inch) cake tin with greaseproof paper, or grease it with butter.

Sift the flour and baking powder together into a bowl. Put the dark chocolate and butter into a heatproof bowl over a pan of simmering water. Allow to melt, then remove from the heat and stir in the sugar. Mix in the eggs and finally fold in the flour mixture, walnuts and chilli chocolate.

Spread the mixture into the cake tin and bake in the oven for 20 minutes. Leave to cool in the tin, then remove and cut into squares.

Serve each brownie topped with a scoop of orange sorbet.

INDEX

ACKNOWLEDGEMENTS

Publisher: Stephanie Jackson
Managing Editor: Clare Churly
Copy-editor: Annie Lee
Art Director: Jonathan Christie
Designer: Jaz Bahra
Illustrators: Abigail Read, Charlotte Strawbridge, James Strawbridge
Photographer: Nick Pope
Stylist: Alison Clarkson
Kitchen Dogsbody: Jim Tomson
Senior Production Controller: Lucy Carter

Picture credits

All photographs © **Nick Pope** with the exception of the following: **Fotolia**/Monica Butnaru (used throughout). **Strawbridge Family Archive** 20 (left). **Thinkstock/** iStockphoto (used throughout).

Illustrations:
Abigail Read 41, 53, 56, 63, 94, 97, 110.
Charlotte Strawbridge 16, 28, 48, 122, 154.
James Strawbridge 39.

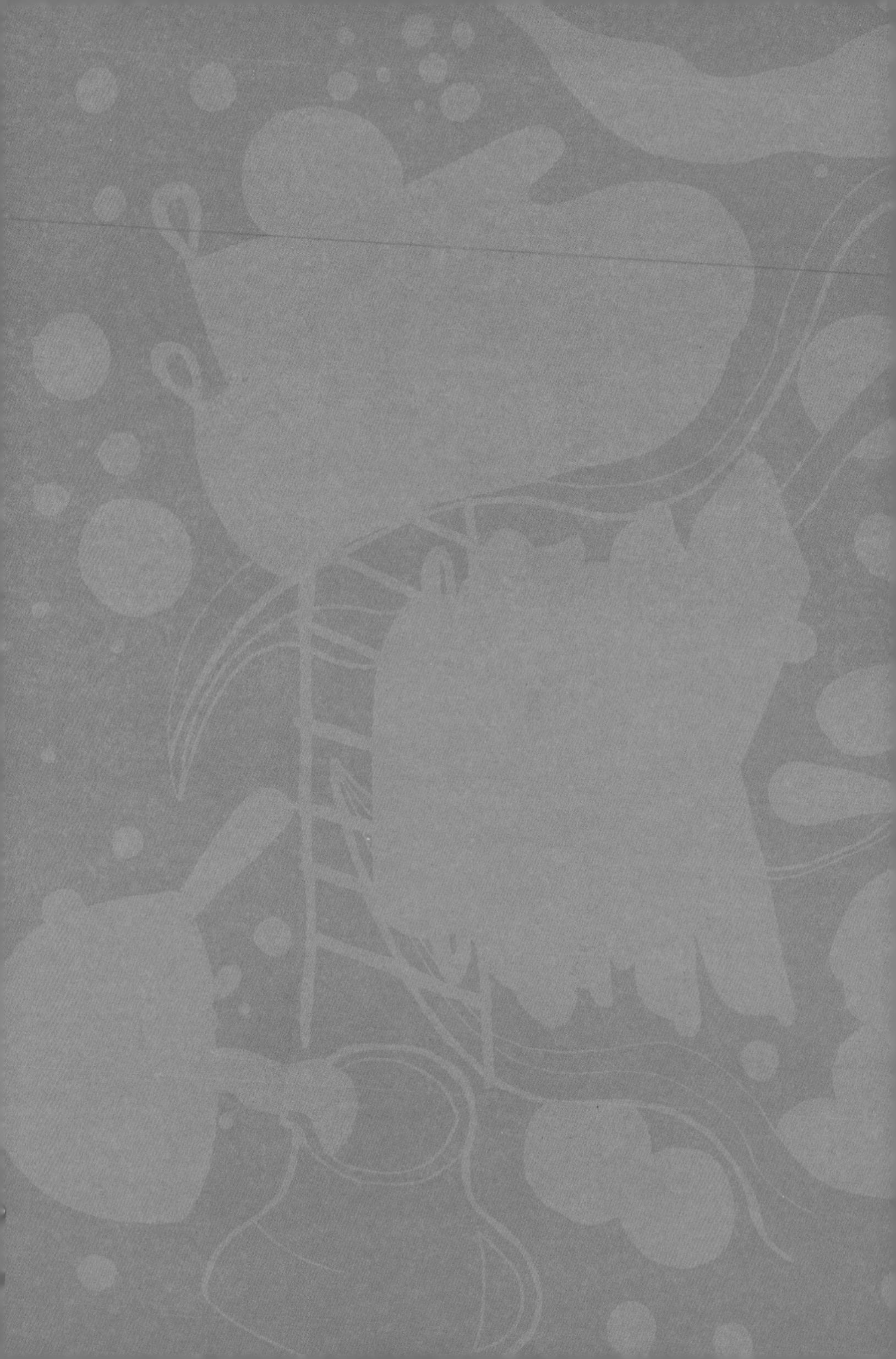